Upstream Engineering and Flow Assurance Forum 2015

Core Programming Area at the 2015 AIChE Annual Meeting

Salt Lake City, Utah, USA
8-13 November 2015

ISBN: 978-1-5108-1876-7

Printed from e-media with permission by:

Curran Associates, Inc.
57 Morehouse Lane
Red Hook, NY 12571

Some format issues inherent in the e-media version may also appear in this print version.

Printed by Curran Associates, Inc. (2016)

AIChE
120 Wall Street, FL 23
New York, NY 10005-4020

Phone: (800) 242-4363
Fax: (203) 775-5177

www.aiche.org

Additional copies of this publication are available from:

Curran Associates, Inc.
57 Morehouse Lane
Red Hook, NY 12571 USA
Phone: 845-758-0400
Fax: 845-758-2634
Email: curran@proceedings.com
Web: www.proceedings.com

TABLE OF CONTENTS

437733 Introduction to Inorganic Scale in Oil and Gas Production

Abstract:

Scale deposits are undesirable in oil and gas production and can cause simultaneous blockage of the production line and dramatic decrease of yield. Oil and gas field scale formation and control becomes even more important in ultra-high temperature and pressure offshore fields due to the enormous costs of remediation and in unconventional production due to the marginal costs of drilling and operation. From the reservoir to the surface pressure can change by as much as 20,000 psi, temperature by 300 F, and brine composition by ± 300,000 mg/L TDS due to water evaporation or condensation in low water cut wells. Sampling and analysis protocols are needed for semi-continuous data collection remotely. The solubility of calcite and barite at most T, P, and composition values is probably as good as is needed, but for few other scale types are similarly robust understanding available, e.g. halite, iron oxides and sulfides, calcium and strontium sulfate, and silicates. Beyond descriptive solubility, there is a dearth of nucleation, growth, and dissolution rates at realistic field conditions. Calcite and barite scale formation are typically prevented by adding milligram per liter of threshold inhibitors, but very little is known about inhibition of most other scale types. Produced waters are naturally anoxic, but testing scale types formed from redox sensitive elements (mostly iron and sulfur) is rarely accomplished due to the difficulty of conducting experiments with strictly anoxic waters (O_2 <<1 mg/L, ppb). This presentation will review the state of knowledge in scale formation, equilibrium modeling, a few kinetic results that appear reliable, and present recent lab and field testing at strictly anoxic conditions. And finally this research will posit a priority list of research needs for economic scale control.

437631 Fundamentals of Asphaltene Precipitation and Deposition

Abstract:

The asphaltene fraction of crude oil contains the highest molecular weight and densest components in petroleum. Asphaltenes are known to cause numerous complications throughout petroleum industry, including production, transportation, and refining. When asphaltenes precipitate from the crude oil to form a solid phase, they can plug pores in oil reservoirs, deposit onto production wellbores and foul heat exchangers in refineries. Despite over a century of research, the properties and behavior of the asphaltene fraction of crude oil remains a topic of intense study and debate. This talk will provide an introduction to the properties of asphaltenes and the modern tools that are used to study their phase behavior and transport characteristics, including aggregation and deposition. A review of both modeling and experimental studies will be discussed with the goal of providing a general background about previous work and new opportunities available to study asphaltenes and their impact on the petroleum industry.

437633 Principles of Wax Deposition and Its Impact on Oil Production

Abstract:

Wax deposition is a critical operational challenge to the oil industry. Waxy components in crude oils represent a group of normal alkanes with carbon numbers greater than 20. These components are normally dissolved in oil at elevated temperature and pressure, such as under reservoir conditions. However, as the crude oil leaves the reservoir and travels towards processing facilities, its temperature and pressure can decrease substantially, causing waxy components to precipitate. Precipitation of wax can lead to deposition on the tubing, pipelines and process equipment and thereby posing severe risks on oil production and transportation. As early as in 1928, wax deposition was reported as a "one of the most troublesome problems in the production of crude oils". This talk will discuss the governing chemical engineering principles in wax deposition phenomena as well as how these principles can be applied to understand wax deposition and to provide guidelines to the remediation of waxy deposition risks.

433458 Fundamentals of Emulsion Behavior: Toward an Understanding of Critical Flow Assurance Challenges

Abstract:

While the behavior of oil into water emulsions and dispersions is relatively well understood, the behavior of water into oil emulsions and dispersions is less well grounded in fundamentals, particularly for multiphase transport in oil pipelines, even under well dispersed conditions. Accurate knowledge of water drop size distribution (DSD) is required to design separation equipment and to assess the potential for gas hydrate formation in deep sea pipelines. In turbulent flow, the pumping and power requirements are determine by geometry and the macroscopic properties of the multiphase mixture. However the drop size, particularly for stabilized emulsion and dispersions, is controlled by microscale flow features that exist on the drop scale and depend on local energy dissipation rate and the kinematic viscosity of the carrier phase. In water like flows, oil drops generally find themselves in a well characterized turbulent flow but in oil flows, water drops often find themselves in a laminar like flow field, since their size can be less than that of the smallest turbulent eddies. In this talk, we will introduce the basic features of turbulence on small scales and discuss currently accepted methods for predicting drop size for oil into water dispersions. We will then consider the state of knowledge for predicting drop size for water into oil dispersions. Critical gaps in data and theory will be assessed and the need for further studies will be discussed. Considerable emphasis will be placed on extrapolating data acquired in stirred vessels, static mixers and other contactors to conditions that exist within the flow assurance environment.

437632 Introduction to Gas Hydrate Formation and Mitigation

Abstract:

In the context of oil and gas operations, gas hydrates are crystalline structures that form when water and hydrocarbons come into contact at low temperatures and elevated pressures. The oil and gas industry currently spends millions of dollars annually on the mitigation and risk management associated with hydrates. The formation of these solid structures can lead to pipeline blockages that can result in ruptures and lost production. Avoiding the formation of these solid blockages is critical for maintaining sufficient production rates. Though hydrates have been researched for decades, there still remain significant technical challenges in managing and preventing hydrate blockage formation in upstream oil and gas operations. This presentation will focus on the background of gas hydrates, why they represent a significant challenge for the oil and gas industry, and recent advances toward improving our understanding and management of hydrates in upstream applications.

431812 Two Dimensional Acoustic Mapping of Fluid Transport in Porous Media – Oil Production Technology Development Applications

Abstract:

Two Dimensional acoustic mapping of fluid transport in porous media – oil production technology development applications

Marc Cassiede and **John M. Shaw**, Department of Chemical and Materials Engineering, University of Alberta, Edmonton, Canada

Abstract. Existing and proposed process concepts based on hydrocarbon and non-hydrocarbon injection into hydrocarbon reservoirs for enhancing production impose a need for better understanding of diffusive mass transfer phenomena in porous media at the pore scale. A clear understanding of active diffusion mechanisms [1] and accurate mutual diffusion rates [2] underlie the quality of reservoir management and hydrocarbon production models. For example, Athabasca bitumen and other high-mean-molar-mass and structured organic materials exhibit complex phase behaviors at the nano- and micro- meter length scales [3,4]. These fluids comprise well-dispersed nanoscale asphaltene-rich phase domains and maltene-rich phase domains that undergo phase transitions in a largely independent manner. The controlling diffusion mechanism in these fluids was recently shown to be composition dependent [1]. Both Fickian and Single File diffusion mechanisms are active. As these resources often arise within limestone or sandstone formations, the presence of micro and nano channels in these media is expected to impose additional constraints on the composition ranges where one or the other or both diffusion mechanisms are active and on observed diffusion rates. Detailed knowledge of transport processes in these porous media and diffusion between injected fluids and hydrocarbon resources in particular comprise crucial and needed process knowledge gaps. Real time measurement of local composition and flow in petroleum reservoir rocks is challenging. A high resolution acoustic technique based on phased array technology has been developed [5] where two multi-element acoustic sensors are used in transmission mode to record times of flight of acoustic waves that propagate through a medium placed between pairs of transmitter-receiver sensors located along the probes. Automated data acquisition and data processing procedures permit the generation of short time interval two-dimensional speed of sound maps. Illustrative near-pore-scale real-time imaging results showing imbibition front movement and mutual diffusion of light hydrocarbons and Athabasca bitumen in natural porous media are presented and discussed.

[1] M. Alizadehgiashi and J. M. Shaw, Fickian and non-Fickian diffusion in heavy oil + light hydrocarbon mixtures, Energy Fuels, 29 (4), 2177-2189 (2015).

[2] H. Fadaei, J. M. Shaw, D. Sinton, Bitumen-Toluene Mutual Diffusion Coefficients Using Microfluidics, Energy & Fuels, (27) 2042-2048 (2013).

[3] A. Bazyleva, M. Becerra, D. Stratiychuk-Dear, J. M. Shaw, "Phase behavior of Safaniya vacuum residue", Fluid Phase Equilibria, Volume 380, 25, 28-38 (2014).

[4] S. Reza Bagheri, A. Bazyleva, M. R. Gray, W. C. McCaffrey, and J. M. Shaw, Observation of Liquid Crystals in Heavy Petroleum Fractions, Energy & Fuels, 24 (8) 4327–4332 (2010).

[5] M. Cassiède and J. M. Shaw, Non-intrusive, high-resolution, real-time, two-dimensional imaging of multiphase materials using acoustic array sensors, Rev. Sci. Instrum. 86, 044902 (2015).

420810 Effective Modeling of Production Decline in Fractured Condensate Reservoirs

Abstract:

Problem Statement

Gas production from fractured reservoirs can be very prolific, despite the extremely low matrix permeability that may be encountered. Knowing the limitations to production is important for operators who drill into very tight matrix reservoirs. To quantify production, the interaction between flowing hydrocarbons, water, and matrix must be well understood. In condensate reservoirs, where gas-liquid phase transition can significantly reduce production rates, the situation is even more complex.

The traditional approach to productivity assessments of fractured reservoirs includes using a detailed 3D reservoir model with detailed fracture network geometry obtained from separate mechanical modeling and/or field observations using microseismic devices and simplified analytical and semi-analytical equations. Although the detailed 3D analysis is both time- and resource-consuming and requires highly qualified personnel to run, the simplified analytical equations are not applicable for evaluating the condensate field production.

Objectives and Scope of Study

The primary goal of this work is to develop an application that will quickly predict pressure depletion and production decline of fracture-stimulated condensate reservoirs. For this purpose, an efficient numerical model was built, based on simplified parallel-planes geometry, but using a full three-phase transient analysis of pressure distribution in an extended stimulated domain to account for gas condensation and water flow. The effectiveness of the approach enables the use of this model as an application to existing wellbore simulators, reducing the need to combine them with a detailed reservoir simulator when a fast parametric or sensitivity analysis is needed.

Method

In the effective approach to the problem, the computation domain is defined as half-space between two parallel fractures and extends from the wellbore to the boundary of reservoir. The model solves for gas, oil, and water flow parameters, and accounts for gas-oil phase transition. To solve the corresponding transient equations, the alternating direction iterations (ADI) have been used. The condensation/evaporation process is simulated using the PVT tables, which are downloaded before the simulations begin. It was assumed that the reservoir initially contained only dry gas and water, and the oil phase was produced during the pressure depletion in the vicinity of the fractures. Because of significant fracture lengths, the pressure distribution inside of each fracture is calculated by solving the lubrication theory equations. This solution is dynamically combined with a solution outside of the fractures; consequently, the pressure profile in the fractures is updated at every time step.

Results and Observations

The developed numerical model has been realized in MATLAB code and used for sensitivity analysis of reservoir productivity regarding changes of fracture size and spacing, as well as reservoir permeability. Because the model enabled phase transition between oil and gas, the banking effect (production reduction attributable to gas condensation) was identified and analyzed.

Conclusions

The obtained numerical model enables the prediction of the production decline in the fractured condensate reservoirs with a detailed account for multiphase reservoir flows and reservoir properties. The simplicity of the fracture geometry used makes the simulations fast and model usable in the form of application for a wellbore solver.

Applications

The developed mathematical and numerical model can be used to perform approximate production decline analysis with a detailed account for condensate properties, including the phase transitions. The simplicity of the model makes it possible to run this analysis in a form of an application for a wellbore simulator, reducing the need to combine it with reservoir solvers.

Innovations

The pressure field and production rate can be calculated by using simple fracture geometry, but taking into account detailed physical-chemical properties of the condensate and material properties of the reservoir.

425499 Fully Compositional Multi-Scale Reservoir Simulations with Mineral Deposition Using the Coupled AD-Gprs/Gflash System

Abstract:

The Gibbs-Helmholtz constrained (GHC) equation of state (Lucia, 2010; Lucia et al., 2012) is a new equation of state that uses molecular scale information and a novel up-scaling equation to estimate the attraction parameter in the Soave-Redlich-Kwong equation of state (Soave, 1972). Recently, a general purpose modeling framework for simultaneous phase and chemical reaction equilibrium has been developed around the GHC equation of state for systems containing any number of ions (Lucia et al., 2014). This new modeling framework permits molecular salt (or mineral) deposition/dissolution, has been implemented within a FORTRAN library called GFLASH, and has been interfaced with the Automatic Differentiation General Purpose Research Simulator (AD-GPRS) (Voskov and Zhou, 2012).

In this talk, a number of isothermal, isobaric (TP) flash solutions and phase diagrams for systems containing multiple dissolved ions will be presented and validated using comparisons to experimental data from the open literature. Finally, numerical results from fully compositional simulations of CO2 injection into the Johansen formation with multiple fluid phases and dissolution/precipitation of multiple salt phases will be presented and discussed.

425830 Reversibility of Asphaltene Precipitation Considering Kinetic Effect Using Temperature-Induced Aggregation

Abstract:

Previous work has shown that asphaltene precipitation is a reversible process using pressure and composition to destabilize asphaltenes, but demonstrated only partial reversibility using temperature changes. Recently, it has been shown that slow kinetic effects play a critical role in the accurate monitoring of asphaltene precipitation, which has not been previously considered in reversibility studies. Consequently, there is an overarching need to investigate temperature-induced asphaltene aggregation with specific monitoring of the kinetic effect. In this work, kinetic asphaltene precipitation was investigated using temperature changes to induce asphaltene aggregation. The amount of precipitated asphaltenes as a function of time at different temperatures was obtained. It can be explicitly stated that the solubility of asphaltenes decreases with decreasing temperature. A phase diagram for asphaltene solubility as a function of solvent quality and temperature was created. In addition, temperature cycling experiments were performed to investigate the reversibility of asphaltene precipitation and thermodynamic models were applied to the results.

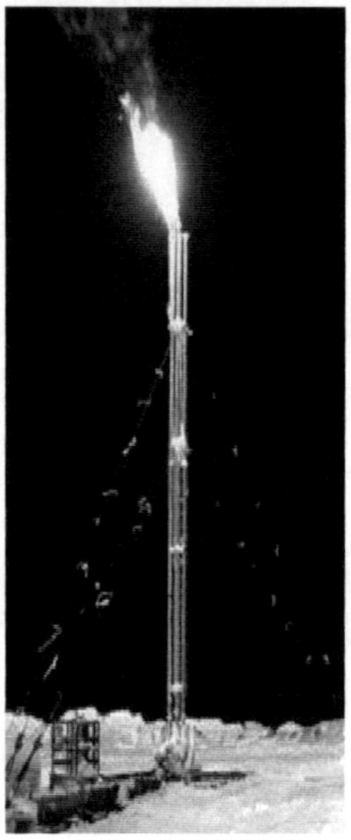

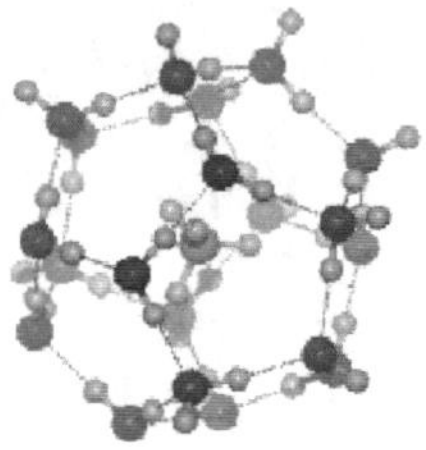

3D Heterogeneous Reservoir Modeling Forecasts of Gas Production by Depressurization of the Prudhoe Bay Unit "L Pad" Hydrate Deposits

Taiwo Ajayi,[1,2] Brian Anderson,[1,2] Evgneniy Myshakin,[1,3]

[1]U.S. Dept of Energy National Energy Technology Laboratory, Morgantown, WV
[2]Department of Chemical Engineering, West Virginia University, Morgantown, WV
[3]AECOM, Pittsburgh, PA

West Virginia University
Department of Chemical Engineering

NETL-RUA
Regional University Alliance

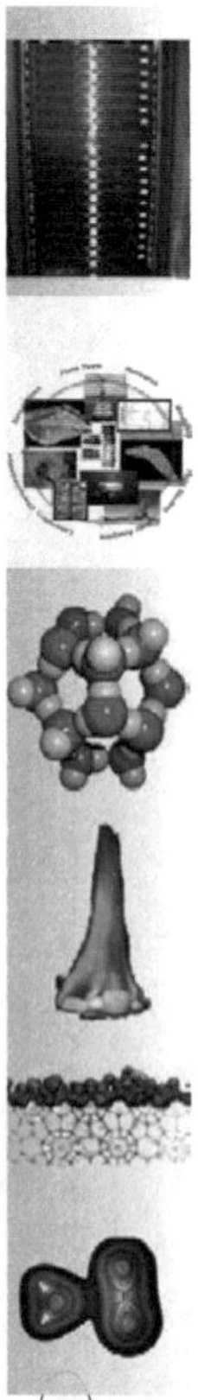

Outline

- **INTRODUCTION**
  - Motivation and Objectives
  - Structure of Hydrates
  - Stability of Hydrates
  - Key Terms
- **METHODS**
  - Study Area
  - Field Data
  - Model Characterization Workflow
  - Numerical Flow Simulation
- **RESULTS**
- **SIGNIFICANCE OF WORK**
- **CHALLENGES AND FUTURE WORK**

INTRODUCTION

Motivation and Objectives

- Provide support to a proposed short-term field production test to be carried out on the Alaska North Slope (ANS)
- Large research efforts in the US, Japan, South Korea, India and China for long term production
- Obtain advanced evaluation of the gas production potential of hydrates as an energy resource
- Study reservoir response to perturbation by depressurization at constant pressure
- Obtain gas and water production rate predictions and identify the key drivers of reservoir performance
- Make recommendations on ways to enhance productivity and economics of gas hydrate reservoirs

Structure of Hydrates

- Molecules are not chemically bound to the water molecules but instead are trapped within their crystalline lattice

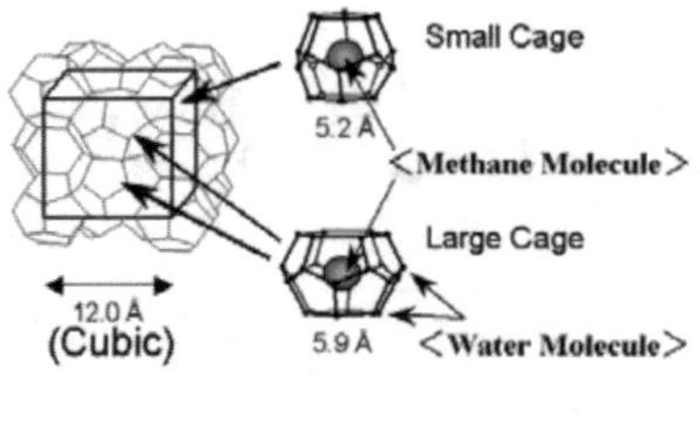

Image Source: ERC Kitami Institute of Tech.

- With the availability of free water and hydrate-forming gas, they form under conditions of high pressure and low temperature
- 1 m^3 of hydrate contains about 160 m^3 of gas

Stability of Hydrates

- Exposure to pressure and temperature conditions outside its stability region makes it to dissociate
- Other factors which affect hydrate stability are salinity and presence of heavier hydrocarbons as guest molecules
- Recovery methods include depressurization, thermal stimulation and hydrate inhibition

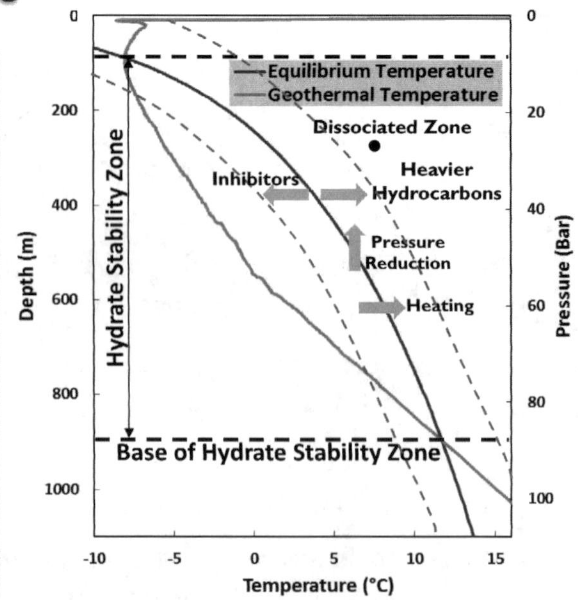

Hydrate Stability in the ANS Permafrost

METHODS

Study Area

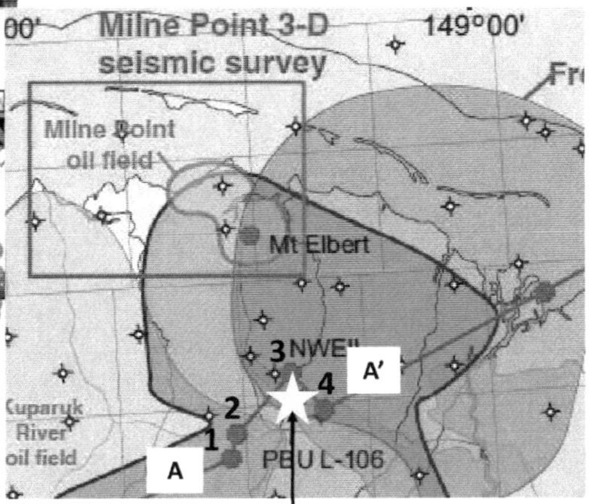

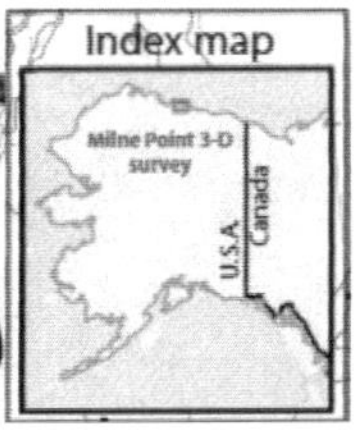

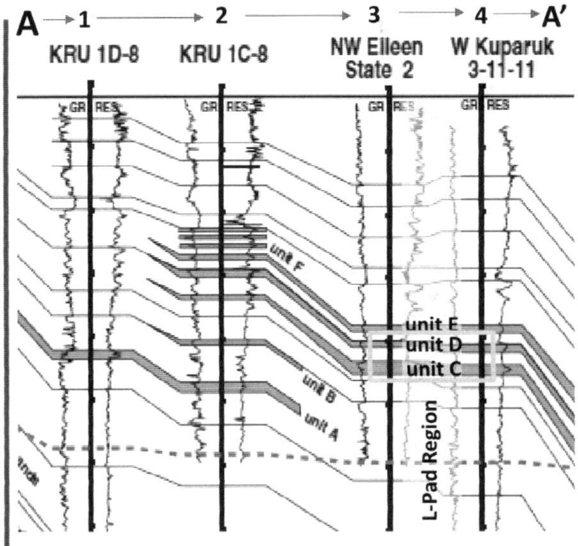

Map of Hydrate Accumulations and Well Cross Sections in the Prudhoe "L-Pad" Region (Collett, 1993)

Field Data (USGS)

- Contour data of the depth to top of Hydrate bearing sand
- D and C sand top picks for 78 wells
- Volume of shale (Vsh) log data from 78 wells
- Resistivity-derived hydrate saturation log data (Sh) from two wells (L-106 and Ignik Sikumi)
- Density logs from two wells (L-106 and L-112)

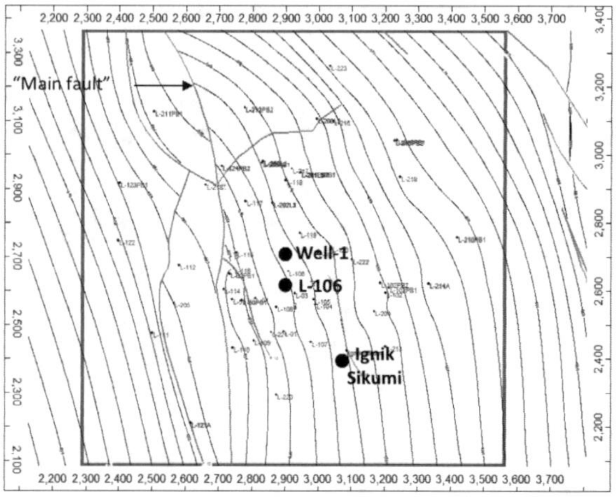

Contour Map of Depth to Top of D Sand

Model Characterization Workflow

I — **Developed Shale Volume-Porosity Correlation**
- Used log data from two wells L-106 and L-112
- Porosity is derived from density log

II — **Calculated porosity values in other 76 wells**
- Porosity was calculated using the developed Vsh-por correlation
- Depth interval is within 100 m from the top of the D sand

III — **Developed porosity variogram models**
- Measure of the variability of porosity values with respect to spatial coordinates

IV — **Performed Gaussian simulation + kriging**
- Gaussian simulation ensures original data points are honored
- Multiple realizations were developed

Vsh-Por Correlation

Conversion of density to porosity in wells L-106 and L-112 were done using the following relations[1]:

$$\varphi = \frac{\rho - \rho_m}{\rho_w - \rho_m}, \qquad \varphi_{sh} = \frac{\rho_{sh} - \rho_m}{\rho_w - \rho_m}, \qquad \varphi_c = \varphi - \varphi_{sh} V_{sh}$$

ρ=density log value

ρ_m=matrix rock density=2.65

ρ_w=density of water=1.00

ρ_{sh}=shale density≅2.37

φ=porosity without correction

φ_{sh}=shale porosity correction

φ_c=shale–corrected porosity

V_{sh}=volume of shale from log

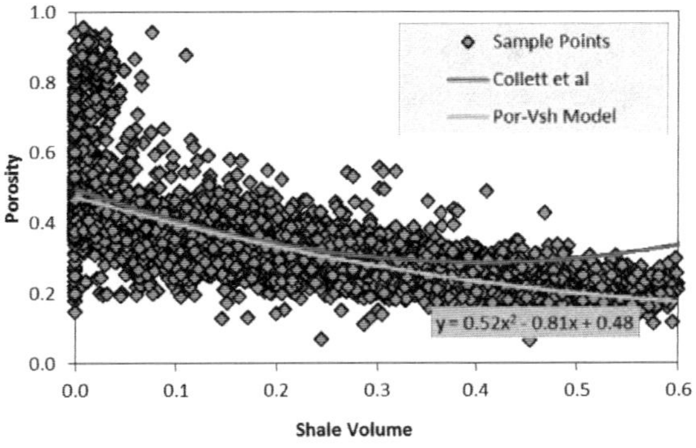

Pooled Porosity-Shale Volume Correlation from L106 and L112 Wells Within the Depth Interval (33 – 853 m)

[1] **Crain, E.R.** Porosity from density log model. *Crain's Petrophysical Handbook.* [Online] 2015. https://www.spec2000.net/12-phidn.htm

Sand Unit Intervals

- D sand interval 0 – 20 m (20 m thick)
- DC intermediate shale 20 – 50 m (30 m thick)
- C sand interval 50 – 100 m (50 m thick)
- Three separate geostatistical analyses were performed for each of the units
- Range of calculated porosity is 0.16 – 0.48

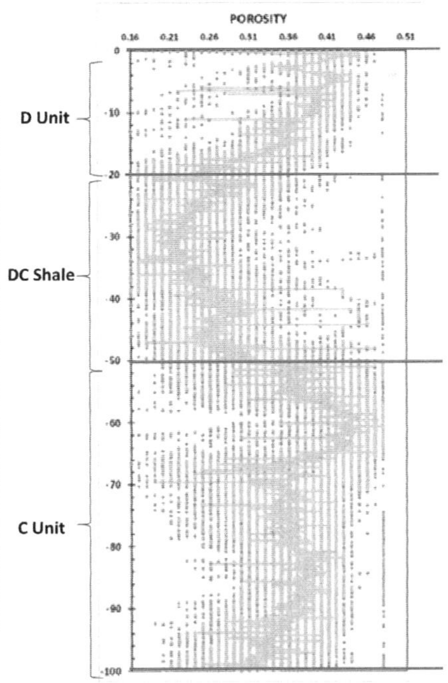

"L-Pad" 3D Reservoir Model

Porosity Distribution

Hydrate Saturation Distribution

Hydrate Saturation Distribution X-Section

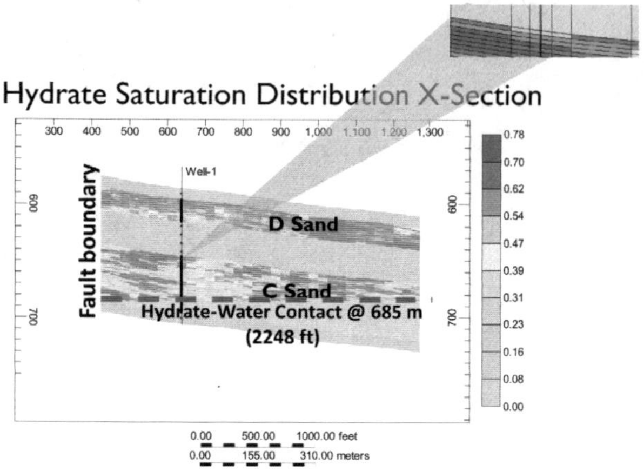

- Figure shows 3D heterogeneous distribution of porosity in the D and C sands with low porosity-overburden, intermediate and underburden shale formations
- 3D heterogeneous distribution of porosity is linearly translated to hydrate saturation with S_H in the shale set to zero
- Vertical cross section shows region to the east of the main fault with hydrate water contact at 685 m depth
- Vertical grid refinements in the HBS and horizontal refinements near wellbore
- Well is optimally located production is operated at 2.8 MPa

Initial Reservoir Conditions and Parameters

Parameters	Hydrate Zone		Shale Zone	Data Source / Reference
	D Sand	C Sand		
Initial (Equilibrated) Temperature (°C)	4.7 – 5.4	6.3 – 8.0	Top: 2.9 – 4.6 Bottom: 8.0 – 10	Ignik Sikumi Temperature Well Log
Intrinsic Permeability	3 - 1643 mD (x,y direction) 1.5 – 822 mD (z direction)		0.001 mD (x,y direction) 0.0001 mD (z direction)	Permeability-porosity correlation
Water Relative Permeability	$k_{rw} = \left\{\frac{(S_w - S_{wir})}{(1 - S_{wir})}\right\}^{5.04}$			Stone (1970) Gaddipati (2008) Anderson et al (2011)
Gas Relative Permeability	$k_{rg} = \left\{\frac{(S_G - S_{Gir})}{(1 - S_{Gir})}\right\}^{3.16}$			
Irreducible Water Saturation (S_{Wir})	0.10		0.30	
Irreducible Gas Saturation (S_{Gir})	0		0	
Capillary Pressure	$P_{cap} = -P\left[(S^*)^{-1/\lambda} - 1\right]^{1-\lambda}; \; S^* = \frac{S_W - S_{Wir}}{1 - S_{Wir}};$ $\lambda = 0.7744$			Van Genuchten (1980)
Pore Compressibility (kPa^{-1})	5.0×10^{-7}			Anderson et al (2011)
Rock Density (kg/m^3)	2650			
Rock Thermal Conductivity (W/mK)	2			

14

NATIONAL ENERGY TECHNOLOGY LABORATORY

WestVirginiaUniversity.
Department of Chemical Engineering

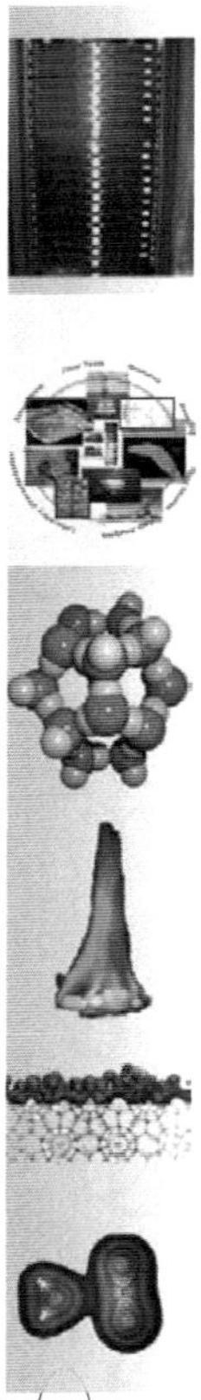

Numerical Flow Simulation: CMG STARS

- STARS simulates non-isothermal reservoir processes
- Hydrate is modeled as the only component of a highly viscous black oil phase
- Uses an equilibrium model for hydrate dissociation
- Reservoir is discretized into grid blocks (cells) and finite difference forms of three mass conservation equations, one energy equation and one phase constraint equation are solved simultaneously for each grid block
- Newton-Raphson method for a system of equations is used

RESULTS

Rates and Cumulative Volumes

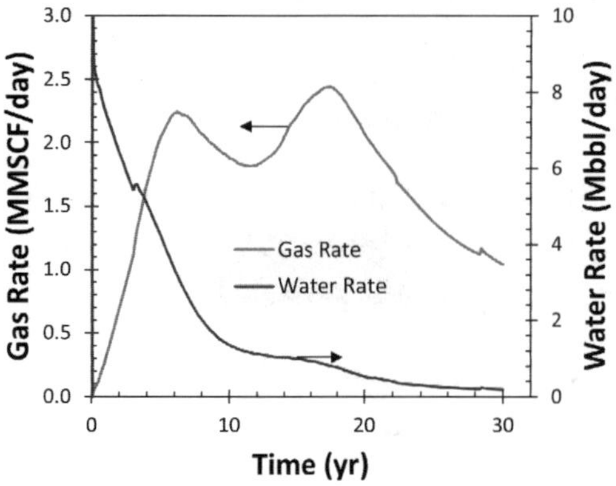

Gas and Water Rate Profiles

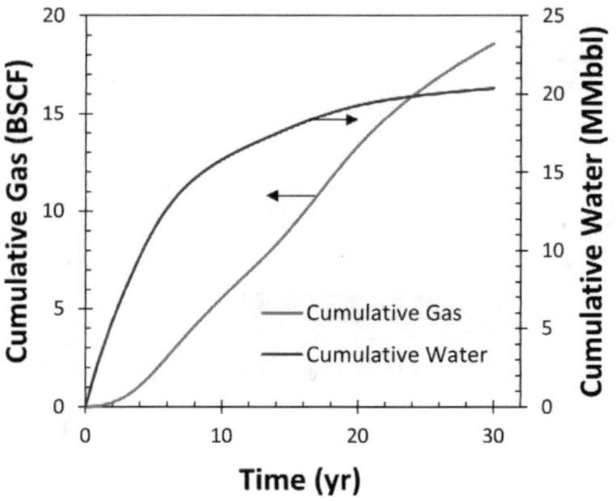

Cumulative Gas and Water Volumes

NATIONAL ENERGY TECHNOLOGY LABORATORY
West Virginia University
Department of Chemical Engineering

Evolution of Hydrate Saturation

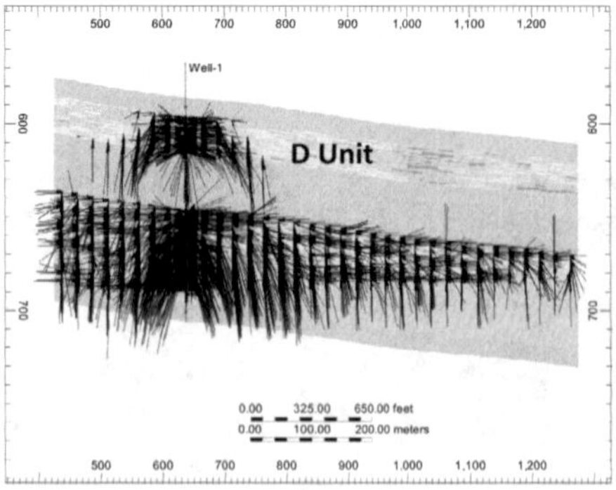

Vertical Cross Section through the wellbore showing hydrate saturation distribution after 4 years

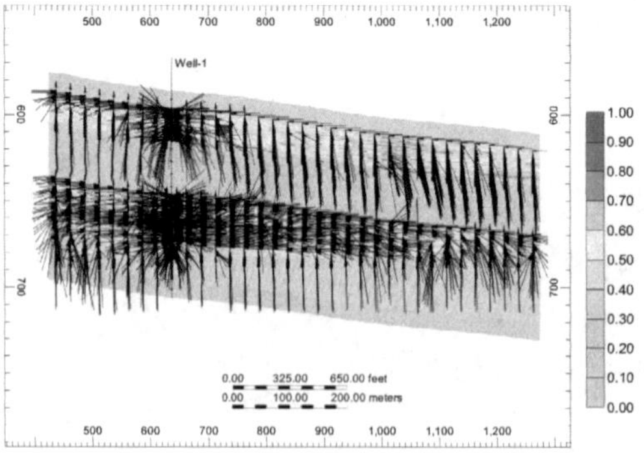

Vertical Cross Section through the wellbore showing hydrate saturation distribution after 15 years

NATIONAL ENERGY TECHNOLOGY LABORATORY
West Virginia University
Department of Chemical Engineering

Evolution of Reservoir Pressure and Temperature

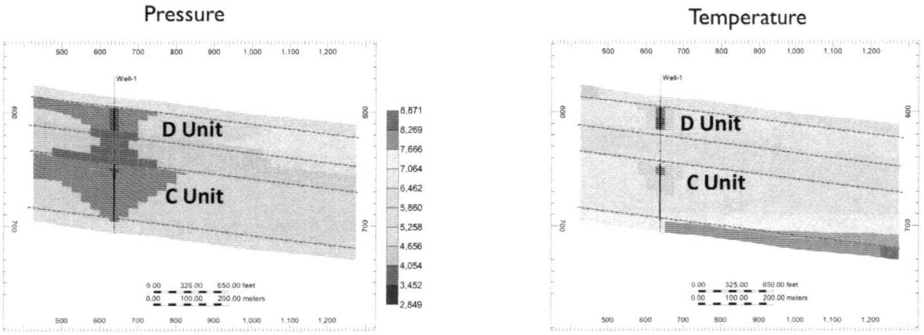

Pressure and Temperature distribution after 4 years

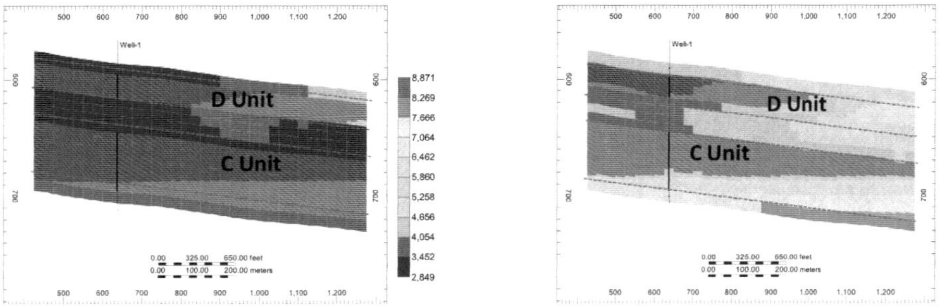

Pressure and Temperature distribution after 15 years

SIGNIFICANCE OF WORK

- Predictions so far have provided a sense of the order of magnitude of gas production rate that could be achieved from a realistic hydrate reservoir model
- Model has shown that effects of secondary hydrate formation in the reservoir on gas production are minimal
- It has also set a basis for the investigation of possible production schemes to improve on gas recovery

CHALLENGES AND FUTURE WORK

- Lack of field production data for continual validation and refinement of reservoir models
- Secondary hydrate and ice formation during production which lead to flow assurance problems
- Significant concerns about possible subsidence of the hydrate-bearing rock as hydrates are being dissociated
- A clearer understanding and incorporation of the kinetics of hydrate dissociation is still required to justify (or not) approximations by equilibrium models

ACKNOWLEDGEMENTS

- **Funding** - Department of Energy (DoE), USGS
- **Data** - United States Geological Survey (USGS)
- **Research Advisor** - Dr Brian Anderson
- **NETL/DoE Supervisor** – Dr Evgeniy Myshakin
- **Research Colleagues** - Manish Nandanwar, Prathyusha Sridhara

REFERENCES

- **Ajayi, T., Anderson, B., Evgeniy, M. and Seol, Y.** Numerical Simulations of Depressurization-Induced Gas Production from a 3D Heterogeneous and Anisotropic Reservoir Model of "L-Pad" Hydrates Prudhoe Bay Unit, Alaska North Slope. To be submitted.
- **Sloan, Dendy E. and Koh, Carolyn A.** *Clathrate Hydrates of Natural Gases*. Boca Raton : CRC Press, 2008.
- **NETL.** *Energy Resource Potential of Methane Hydrate*. The Energy Lab, p. 5., 2009.
- **Collett, T.S., Agena, W.F., Lee, M.W., Zyrianova, M.V., Bird, K.J., Charpentier, R.R., Cook, Troy, Houseknect, D.W., Klett, T.R., Pollastro, R.M., and Schenk, C.J.**, 2008, Assessment of gas hydrate resources on the North Slope, Alaska, 2008: U.S. Geological Survey Fact Sheet 2008-3073, 4 p.
- **Stone, H.L.,** Probability model for estimating three-phase relative permeability. J. Pet. Tech, 1970. 22(2): p. 214-218.
- **Aziz, K. and A. Settari,** Petroleum Reservoir Simulation. 1979: Applied Science Publishers
- **Gaddipati, Manohar**, Code Comparison of Methane Hydrate Reservoir Simulators using CMG STARs. An Unpublished MSc Thesis Submitted to the Department of Chemical Engineering, West Virginia University.

THANK YOU!

Definition of Some Key Terms

- **Rock Porosity** - Fraction of void space in a rock (%)
- **Hydrate Saturation** – Relative amount of hydrate in the pores of the hydrate-bearing rock (%)
- **Irreducible Water Saturation** – Relative amount of water that remains (clay- / capillary-) bound to the reservoir pores during production
- **Rock Intrinsic Permeability** – Ability of a rock to transmit fluid (mD)
- **Effective Permeability** - Ability to preferentially flow or transmit a particular fluid through a rock when other immiscible fluids are present in the reservoir (mD)

$$Q = -\frac{k_{eff}A}{\mu}\frac{\partial p}{\partial x}$$

- **Relative Permeability** - Ratio of effective permeability of a particular fluid at a particular saturation to absolute permeability of that fluid at total saturation (%)
- **Quadruple Point** – Thermodynamic equilibrium state where hydrate, water, ice and hydrate-forming gas coexist

Recovery Methods

- Depressurization
  - Deemed most effective
  - Relatively low energy requirement
- Thermal Stimulation
  - Too expensive to be efficient
  - Secondary recovery technique
- Chemical Inhibitors
  - Also quite expensive
  - Environmental concerns
- Guest Molecule Displacement
  - Novel technique used during the Ignik Sikumi test in 2012 - the first and only field trial so far
  - Pro-green earth!

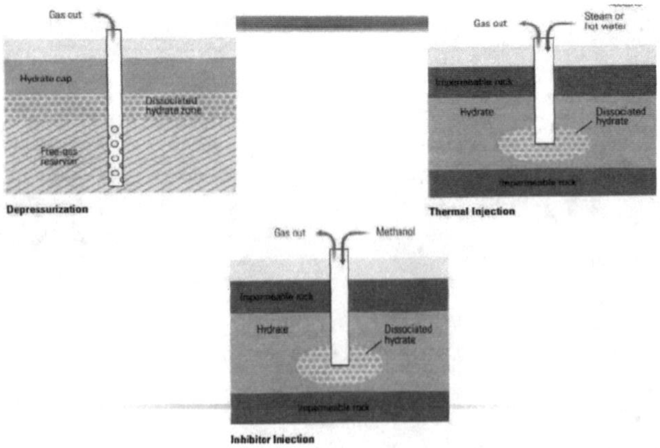

Hydrate Depressurization

I
- Initialized by P_1- P_5
- P_2 = equilibrium pressure at T_1
- Driving force = P_1-P_2

II
- Equilibrium phase change with +ΔH
- T_1 is decreased to T_3 along the equilibrium curve
- Driving force = T_1-T_3 (or ΔH)
- P_3 = equilibrium pressure at T_3

III
- Reaction kinetics of hydrate dissociation
- $\frac{-dn_H}{dt} = k_D A(f_{eq} - f_g)$
- Driving force = $f_{eq} - f_g$ = P_3-P_4

IV
- Fluid flow from dissociation front to depressurized zone
- Driving force = $P_4 - P_5$

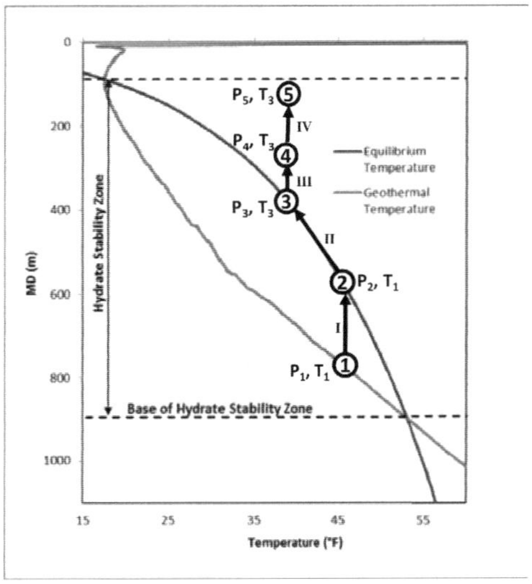

Well Trajectories (approx.)

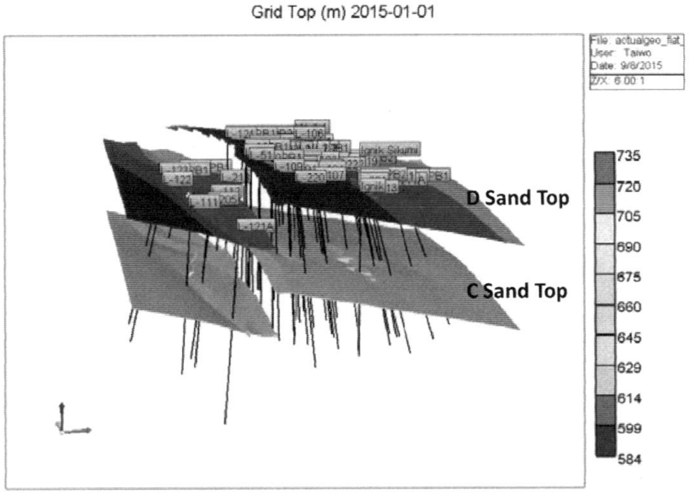

- Area covered by the 78 wells is 1212 x 1281 m.
- The porosity values were calculated along the trajectories of 78 wells penetrating the D and C sand units shown in the figure, at 0.5 ft intervals
- These calculated porosity values were the input values used for the geostatistical analyses of porosity distribution in the entire domain

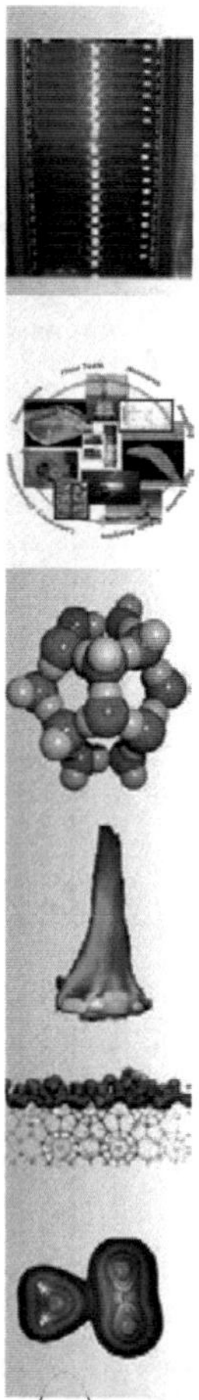

Numerical Flow Simulation: CMG STARS

- STARS simulates non-isothermal reservoir processes
- Hydrate is modeled as the only component of a highly viscous black oil phase
- $CH_4.N_H H_2O_{(h)} \rightarrow CH_{4(g)} + (N_H)\, H_2O_{(w)}, \qquad N_H = 6.176$
- Hydrate dissociation rate is given by

$$-\frac{dn_H}{dt} = \frac{k_d^0 A_{SH} a}{\rho_w \rho_H} \exp\left(-\frac{E_H}{RT}\right)(\varphi S_w \rho_w)(\varphi S_H \rho_H)\left(1 - {}^{1}/_{K}\right)$$

- Reservoir is discretized into grid blocks (cells) and finite difference forms of three mass conservation equations, one energy equation and one phase constraint equation are solved simultaneously for each grid block
- Newton-Raphson method for a system of equations is used

$$X^{k+1} = X^k - [J^k]^{-1}.R^k$$

- $R(X) = [net\ inflow\ rate] + [net\ source/sink\ rate] - [rate\ of\ accumulation]$
- Solution convergence requires that both $X^{k+1} - X^k$ and R are sufficiently small

29

NATIONAL ENERGY TECHNOLOGY LABORATORY

West Virginia University
Department of Chemical Engineering

437831 The Interaction of Synthetic Porphyrins with Asphaltenes: Role in Self Assembly

Monday, November 9, 2015: 2:10 PM
252A/B (Salt Palace Convention Center)

McKay Rytting, University of Notre Dame, Notre Dame, IN and Peter Kilpatrick, Chemical and Biomolecular Engineering, University of Notre Dame, Notre Dame, IN

The Interaction of Synthetic Porphyrins with Asphaltenes: Role in Self Assembly

McKay Rytting and Peter Kilpatrick

Department of Chemical and Biomolecular Engineering

University of Notre Dame

Notre Dame, IN 46556

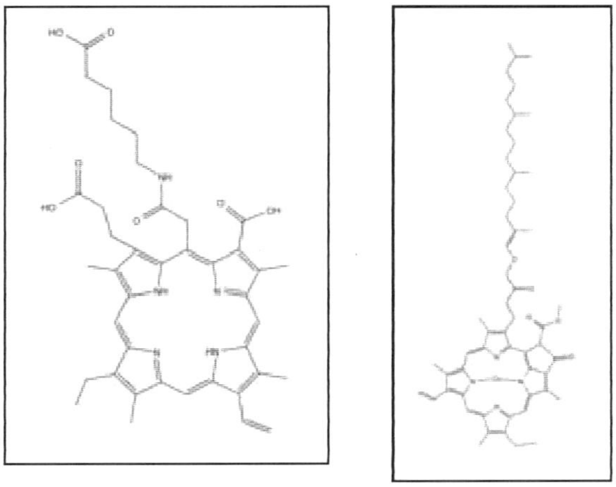

Figure SEQ Figure * ARABIC 1:

Cu(II) Pheophytin a Chlorin e6 6-NH2 Hexanoic Acid

(P10299) Amide (C40286)

Figure 1:
Cu(II) Pheophytin a Chlorin e6 6-NH2 Hexanoic Acid
(P10299) Amide (C40286)

Asphaltenes are known to aggregate at very low concentrations to form nanoscale aggregates that are oblate cylindrical in shape (disk-shaped) and that have characteristic disk diameters in the 2-50 nm size range. Aggregate sizes vary with temperature, solvent, asphaltene chemistry, and with added solvating agents such as demulsifiers and surface-active agents. These aggregates also adsorb onto interfaces (oil-air, oil-water, oil-solid) and molecularly rearrange to form elastic solid films that stabilize foams and water-in-oil emulsions, as well as form deposits during subsea production. Naturally occurring asymmetric petroporphyrins are known to alter asphaltene aggregation and modify the size and interfacial activity of asphaltenes. In an effort to understand this solvating process better, we have performed interfacial tension and rheology experiments, small angle neutron scattering (SANS), and pulsed field gradient spin echo (PFGSE) diffusion NMR experiments on asphaltenes to which two very different synthetic porphyrins (see Figure 1) have been added: one in which there are multiple carboxylic acid functional groups grafted to the exocyclic porphyrin ring and one in which a long branched alkyl tail has been grafted to the exocyclic ring. As one might anticipate, the results are dramatically different. The first synthetic porphyrin with the grafted branched alkyl tail (P10299) solvates the asphaltenic aggregates and reduces aggregate size and molecular weight at all concentrations (see Figure 2). The second synthetic porphyrin with the multiple carboxylic acid grafts (C40286) appears to incorporate into the aggregates and increase aggregate size with increasing concentration at low temperature, but solvates and decreases the size of aggregates at higher temperature (see Figure 3). Results from PFGSE-NMR diffusion measurements show similar trends. These results offer valuable clues into the precise molecular mechanisms of asphaltene self assembly. In this talk, we will present comprehensive data from a battery of experiments, including interfacial tensiometry and oscillating drop dilatational rheometry, as well as the SANS and PFGSE-NMR results.

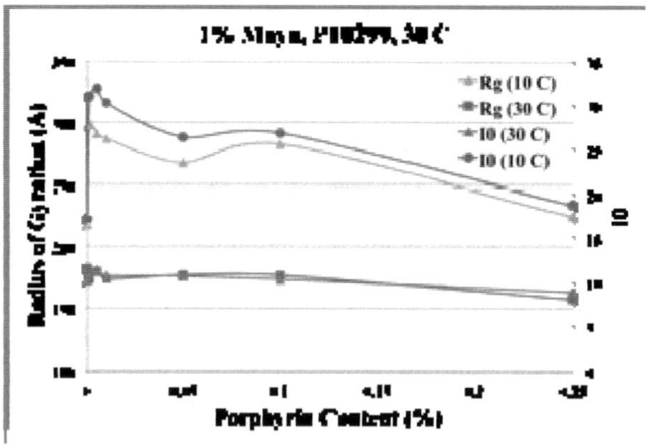

Figure SEQ Figure * ARABIC 2: Effect of P10299 on size and weight of Maya asphaltenic aggregates

420169 Nanoparticles Catalyze Rapid Hydrolysis of Poly(lactic acid) in Downhole Conditions

Abstract:

Poly(lactic acid) (PLA), a bio-based degradable polymer, has been used in the oilfield service industry for multistage hydraulic fracturing. The rate of PLA degradation depends on the rate of ester hydrolysis, which is significantly reduced below the glass transition temperature (T_g) of PLA. This limits the applications of PLA in downhole formations (type of rocks) to only high temperatures. A simple, ready-to-scale-up chemistry using zinc oxide nanoparticles to catalyze the hydrolysis of PLA results in rapid degradation of PLA and extends its applications to low-temperature shale formations. Furthermore, for the first time, we have applied a nondestructive analytical method of ^1H T_2 NMR relaxometry to measure the apparent rate constants of PLA hydrolysis in solid, heterogeneous, composite systems that have multiple and complex reaction kinetics. The demonstrated activation energy for ZnO-catalyzed PLA hydrolysis is about 38% lower than that of pure PLA hydrolysis.

432011 Phase Transition, Fracture Parameters, Relative Permeability, and Capillary Pressure Effects on Condensate Production in Fractured Reservoirs

Abstract:

Problem Statement

Since the commercial production of natural gas began in the United States in 1821 [1], petroleum engineers have been trying to improve productivity by better understanding shale formations, especially since the 1970s. Because of extremely low matrix permeability in shale reservoirs, profitable commercial production currently involves hydraulic fracture networks. Understanding the effect of hydraulic fracturing parameters is therefore beneficial to operators in terms of both fracture design optimization and production enhancement.

Recent fast progress in hydraulic fracturing technology has brought much attention to gas reservoirs, where condensate forming and extraction become problematic. Liquid flow meets significantly more resistance compared to gas in shale fractures filled with proppants and in wellbores. Reservoir productivity also becomes more complex by including condensate phase relative permeability and capillary pressure effects. At the same time, many widely used analytical production decline models work poorly in cases of liquid-rich formations. To perform fracture design optimization and sensitivity analysis, an approach is required that is much less time consuming than detailed reservoir simulation while being sufficiently accurate to capture the physics of the process. To address situations involving gas condensates, it must rely on the numerical simulation of multiphase flow in the interconnected system of the wellbore and fractures within a reservoir, represented by productivity.

An effective numerical model reported in an accompanying paper [2] was developed to predict production decline in a fracture-stimulated liquid-rich gas field. It allows for easy investigation of the impact of gas-condensate system phase transition, fracture parameters, relative permeabilities, and capillary pressures on the well productivity. The model solves the transient pressure depletion process with focus on the effects of condensate banking and condensate-phase-related permeability and capillary pressure changes.

Method

A detailed description of the model calculating transient pressure depletion in the stimulated volume between and around parallel-plane fractures perpendicular to a horizontal wellbore is provided in [2].

The flows of gas, condensate (oil), and water phases are simulated in the model. To describe the gas-oil phase transition in the system, the Extended Black Oil model is used. The gas and oil components can transfer between the hydrocarbon phases according to pressure-volume-temperature (PVT) rules, while the water component remains in the aqueous phase. Calculation of transient mass transport and phase distribution is coupled with pressure field distribution.

Relative permeabilities and capillary pressures are accounted for each phase as functions of phase saturations. With the oil phase having much higher viscosity and resistance while flowing within a tight formation, its banking (blocking) effect on the production rate was simulated and studied in detail.

A sensitivity screening of fracture parameters, including fracture length, fracture spacing, and fracture conductivity was conducted to facilitate fracture design optimization and production enhancement, with the primary focus on production rate and net present value (NPV).

Results

An effective production model of a fractured reservoir was developed for this study. In this paper, the oil condensation process is shown, and the condensation banking effect on the production rate decline is identified and analyzed. The relative permeability and capillary pressure effect of the condensate phase on production is also discussed. Sensitivity analysis of reservoir productivity related to fracturing parameters, with respect to changes of fracture length, fracture spacing, and fracture conductivity, was performed.

Conclusions

To simulate the condensate production decline in fracture-stimulated condensate reservoirs, an effective numerical model was developed. It accounts for a variety of physical effects influencing liquid-rich reservoir productivity, including gas-condensate phase transition, relative permeabilities, and phase capillary pressures. The simplicity and simulation speed of the model makes it possible to incorporate it into a wellbore simulator to model the transient pressure drop and phase transitions in the near-wellbore region, reducing the necessity for more complex reservoir simulations. It also opens up opportunities for applications related to fast fracture and well design optimization and net profit estimates.

References

[1] J.B. Curtis, 2002, Fractured Shale-Gas Systems. AAPG Bulletin, v. 86, No. 11, P1921-1938.

[2] A. Filippov, X. Jia, and V. Khoriakov, 2015, Effective Modeling of Production Decline in Fractured Condensate Reservoirs. Presented at the AIChE Annual Meeting, Salt Lake City, Utah, 8–13 November.

415460 A Robust Control Strategy for Surface Pressure Pulsation Minimization in Hydraulic Fracturing

Abstract:

A stimulation job entails high pressure pumping of proppant and sand into a well-bore in order to induce hydraulic fractures in the subsurface formation, typically, along horizontal wells. There may be 10-25 reciprocating pumps connected at the surface and during simultaneous operation these naturally give rise to high pressure pulsations over the piping system. These pressure fluctuations induce acoustic and mechanical resonance that leads to excessive vibration, which in turn can cause considerable wear and damage to the pump and piping network, potentially with catastrophic consequences. In this work, we present a robust control strategy designed to prevent this eventuality.

The reciprocating pumps employed in stimulation job frequently operate at pressures in excess of 10,000 psig, with pressure fluctuations in hundreds of psig. In a typical reciprocating pump design, rods connected to a crank drive multiple plungers which are offset in phase. Plungers accelerate between maximum positive and negative velocities in a sinusoidal curve. Subsequently, pressure and flow follow sinusoidal characteristics. The pressure and flow rate variation is mitigated due to combination of flow from multiple (three or five) plungers designed to be out of phase within a multiplex pump. When the piping system comprises elbows, tees, or diameter changes, the pressure pulsations can lead to piping vibrations, a phenomenon termed *acoustic-mechanical*coupling. If the vibration-inducing frequency (or the pump pressure pulse frequency) matches the natural frequencies of the piping system, it induces mechanical resonance; and the vibration forces, stresses, and amplitudes can be excessive.

Transient fluid flow in piping networks leads to another source of acoustic resonance. The pressure pulses from the pumps induce wave-guided acoustic modes in the pipes that travel at the wave speed along the pipe. When these bounce off a reflecting surface (such as a valve or a bend in the pipe) they generate standing waves that may produce resonance. The wave speed is calculated using the known acoustic modes in a fluid-filled pipe, which is dominantly the tube wave but could also include the flexural wave. Resonant conditions are achieved when the pump frequency matches the acoustic natural frequency of the fluid-piping system.

Experimental results based on pressure measurements in the time and frequency domains were used to characterize various pumping systems. In particular, experiments were conducted on a 2-pump system to address the issues of fatigue and premature failure (of piping iron and pumps) caused by pressure pulses and resultant vibrations. Furthermore, to study the transient dynamics of a particular pumping system, and to mitigate excessive pulsations, a 10-pump Flowmaster® CFD simulation model was developed using the experimental data. Starting with a rigorous reciprocating pump model, and using fluid properties from NIST REFPROP, a transient fluid flow model with field scale piping layout and fixtures was developed. This model was incorporated in an optimization framework and used to minimize the peak-to-peak (P2P) pressure differential observable at the missile outlet with active pump speed and phase control. Notably, as simulation-based P2P objective function is highly nonlinear, a robust control strategy was introduced to minimize the observable P2P variation. As a reduction in the P2P pressure pulse leads to a concomitant reduction in vibration, this results in reduced maintenance costs and safer field operation. The control strategy and test results based on a 10-pump simulation model will be presented.

OPTIMIZATION OF HUFF-N-PUFF GAS INJECTION IN SHALE CONDENSATE RESERVOIRS TO IMPROVE LIQUID OIL PRODUCTION

Jie Sheng, James J. Sheng, Texas Tech University, Lubbock, TX

Abstract

Earlier studies show that huff-n-puff injection is preferred to continuous gas flooding to improve liquid oil production in shale gas condensate reservoirs. Compared to gas flooding, huff-n-puff has more operation parameters to optimize so that liquid oil production can be maximized. This paper focuses on the effects of cycle time and soaking time under practical operation conditions and reservoir conditions. The operation conditions include injection and production pressures, injection and production rates. The reservoir conditions include reservoir permeability and initial reservoir pressure.

The simulation results and discussions further show that huff-n-puff injection of produced gases can produce more liquid oil in gas condensate reservoirs than gas flooding or primary depletion. The advantages of huff-n-puff over gas flooding are attributable to high drawdown pressure and oil saturation decrease near the wellbore by evaporation, hence overcoming the condensate blockage and pressure transport problem owing to ultra-low permeability. The simulation results show that an optimum soaking time is so short (3 to 5 days) that it is not necessary to design a soaking time in practice because such short soaking time can be met when changing huff mode to puff mode. Although longer soak time results in reduction of oil recovery, the gas utilization (oil recovered per unit volume of gas injected) increase as soak time gets longer. Similarly, the effect of cycle time is studied. It is found that optimum cycle times are in months in typical shale reservoir conditions. Analysis of simulation data about the changes in pressure, oil saturation and oil viscosity, empirical soak time and cycle time are proposed in this paper.

Introduction

Gas condensate is defined as hydrocarbon liquid dissolved in saturated natural gas that comes out of solution when the pressure drops below the dewpoint. Condensate liquid saturation can build up near a well because of drawdown below the dewpoint pressure, ultimately restricting the flow of oil and gas.

In a shale or tight gas condensate reservoir where the formation permeability is extremely low (nano-Darcy or micro-Darcy), if the reservoir pressure is below the dew point pressure, the pressure difference between the well flowing pressure and reservoir pressure which provides the drive force to produce the gas condensate will be low. Consequently, the resulting total hydrocarbon recovery will be low as well. There are two methods to increase reservoir pressure: gas flooding and huff-n-puff gas injection. In the gas flooding process, gas is injected through an injector, and hydrocarbon is produced from a producer that located in another position. In the huff-n-puff gas injection, gas injection occurs during the huff period, while fluid production takes place in the puff period.

Earlier studies show that huff-n-puff injection can produce more liquid oil in shale gas condensate reservoirs than gas flooding or primary depletion with different reservoir and fluid properties and operation conditions (Sheng, 2015). The advantages of huff-n-puff over gas flooding are early response to gas injection, high drawdown pressure, oil saturation decrease by evaporation, and overcoming the pressure transport problem owing to ultra-low permeability.

Setup of Simulation Model

Sheng (2015) has established the huff-n-puff gas injection model in gas condensate reservoirs. The compositional simulator GEM, developed by CMG is employed to set up the model. The gas condensate composition and related parameters are shown in Table 1(Orangi et al., 2011) and the reservoir and fluid properties are shown in Table 2. In Table 1, P_c, T_c and V_c are critical pressure, critical temperature and critical volume, respectively, and MW is molecular weight. The model grid set up is shown in Figure 1. A fracture of 2-ft width and 46.5 mD is used to represent the real fracture of 0.001 ft and 83,000 mD so that the equivalent fracture conductivity ($k_f w_f$) remains the same value. The fracture length is1000 ft., represented with 55 blocks in the J direction. The formation height and the fracture height are 200 ft, represented with 7 blocks in the K direction. Both the injector and the producer are located in I=11, J=28. The wells are perforated at the top layer (K=1). Furthermore, the block sizes vary in all directions, the closer the block is to the fracture, smaller the block size is (Figure 1).

Table 1 Peng-Robinson EOS Fluid Description of Eagle Ford Condensate (Orangi *et al.*, 2011)

Components	Composition (mole fraction)	P_c (atm)	T_c (°K)	Acentric factor (dimensionless)	MW (g/mole)	V_c (1/mol)
C1	0.65882	45.4	190.6	0.013	16.04	0.099
N2	0.00154	33.5	126	0.04	28.01	0.089
C2	0.08337	48.2	305.4	0.0986	30.07	0.148
C3	0.0467	41.9	369.8	0.1524	44.09	0.203
CO2	0.02686	72.8	304.2	0.225	44.01	0.094
IC4	0.01045	36	408.1	0.1848	58.12	0.263
NC4	0.01825	37.5	425.2	0.201	58.12	0.255
IC5	0.00825	33.4	460.4	0.2223	72.15	0.306
NC5	0.00791	33.3	469.6	0.2539	72.15	0.304
NC6	0.01194	32.46	507.5	0.3007	86.18	0.344
C7+	0.07627	27.8	584.1	0.3673	112	0.446
C11+	0.04551	20.2	692.1	0.5491	175	0.685
C15+	0.00278	17.62	737.5	0.6435	210	0.809
C20+	0.00135	15.39	781	0.7527	250	0.942

Table 2 Reservoir and Fluid Properties in the Huff-n-Puff Model

Initial reservoir pressure	9088 psi
Porosity of shale matrix	0.06
Initial water saturation	0.2
Compressibility of shale	$5\text{X}10^{-6}$ psi^{-1}
Shale matrix permeability	0.0001 md

Reservoir temperature	310 °F
Reservoir thickness	200 ft
Dew point pressure	3988 psi

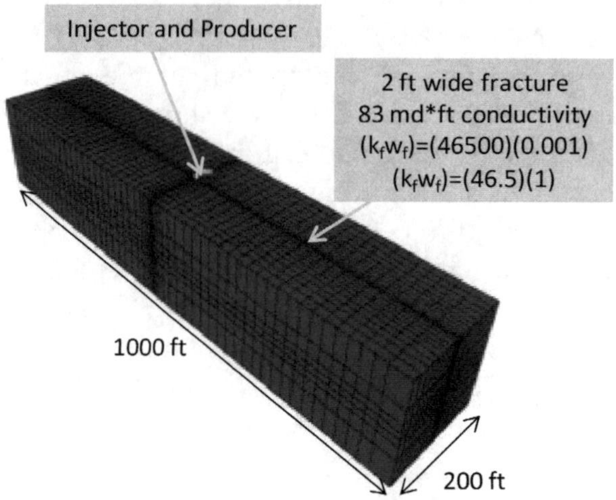

Figure 1 Model Grid Set Up

Optimization Principles

Earlier studies showed that huff-n-puff gas injection can achieve more oil recovery than gas flooding process in shale gas condensate reservoirs. It is imperative to optimize several parameters in the huff-n-puff process. In order to optimize the process, the objective parameter should be defined. Net present value is a reasonable parameter. However, many parameters such as equipment cost, operation cost and oil price are unknown or subject to changes. Therefore, considering the oil price is much higher than gas, oil recovery factor is employed here as an objective parameter.

The parameters that should be optimized in huff-n-puff process include injection time (huff time), production time (puff time) and soaking time. These parameters are related to injection and production pressures, and injection and production rates. Injection and production pressures and rates are mainly determined by the facility constraints. In our model, the initial reservoir pressure is 9088 psi and maximum injection pressure is set to be 9500 psi, which is close to the initial pressure. This is a typical practice for pressure maintenance. The maximum injection rate in the model is set as 300 MSCF/D, 100 MSCF/D and 600 MSCF/D, respectively. Earlier studies (Sheng, 2015) concluded that higher oil recovery is obtained as a lower bottom-hole flowing pressure (BHFP) is used, even though the flowing pressure is lower than the minimum miscible pressure. Accordingly, the minimum production pressure is set as 500 psi. The maximum gas production rate is 300 MSCF/D in the model. The huff-n-puff process initiates after 5 years of primary depletion and lasts for 25 years. In this study, the injected gas is methane.

Optimization Results and Discussions

Based on the above principles, optimum huff time, puff time and soaking time are determined and the number of cycles is also discussed in this paper.

Optimum Huff and Puff Times

The previous literature has not discussed the huff-n-puff gas injection in gas condensate reservoirs. Some studies have investigated the huff-n-puff process in unconventional reservoirs such as tight formations. However, the injection and production times used are very different. Kurtoglu (2013) used 60 days of injection, 10 days of soaking and 120 days of production in her simulation work. Shoaib and Hoffman (2009) used three months in each of injection, soak and production periods. Wang *et al.* (2010) simulated the EOR potential in the tight (0.04-2.5 mD) Bakken formation in Saskatchewan. In their models, one cyclic process includes 10 years of CO_2 injection, 5 years of soaking time, and 5 years of production time.

Table 3 Effect of Huff and Puff Times on Oil Recovery and Gas Recovery (Qg_{max}=300MSCF/D)

Case	Huff (days)	Puff (days)	Oil RF (%)	Gas RF (%)
H100P100	100	100	32.495	30.468
H125P125	125	125	33.141	30.161
H150P150	150	150	34.073	29.711
H180P180	180	180	34.125	29.071
H250P250	250	250	34.532	27.842
H300P300	300	300	34.289	26.62
H500P500	500	500	32.347	31.322
H750P750	750	750	30.42	36.002
H1500P1500	1500	1500	27.556	43.391
H2250P2250	2250	2250	26.147	47.629
H250P250H350P350			34.54	27.491
H250P250H300P300			34.695	26.45

*H250P250H350P350 stands for the case with 4 cycles of H250P250 and 10 cycles of H350P350; whereas H250P250H300P300 stands for the case with 6 cycles of H250P250 and 10 cycles of H300P300.

From the previous study on shale oil of our group, it may be indicated that the optimum huff time is when the near-wellbore block pressure reaches to the set maximum injection pressure, and the optimum puff time is when the near-wellbore block pressure reaches to the set minimum production pressure. To verify the hypothesis, cases with different huff and puff times are investigated. When the maximum gas injection and production rate are set as 300 MSCF/D, the simulation results are shown in Table 3. It is found that among the cases where the huff and puff times maintain constant during the 25 years of huff-n-puff process, the H250P250 is the optimal case for oil recovery. The oil recovery from this case is 34.532%.

Figure 2(a) shows the near-wellbore block pressure in the H100P100 case. In the first five huff-n-puff cycles, the injection pressure reaches over 7000 psi and the production pressure drops to the set minimum production pressure 500 psi. However, the injection pressure in the subsequent cycles gradually decreases to 6500 psi. Also, the production pressure does not continue to be the set minimum value and gradually increases to 1700 psi. The variations of the injection pressure and production pressure are attributable to the hydrocarbon changes in the reservoirs. As more gas condensate is produced, the hydrocarbon left in place tends to be more compressible. In contrast, the near-wellbore block pressure during the huff time in the H250P250 case reaches to the set maximum injection pressure 9500 psi and the near-wellbore block pressure during the puff time reaches to the set minimum production pressure 500 psi, as shown in Figure 2(b). Although the injection pressure gradually decreases to 8800 psi, the drawdown pressure used to produce oil for the H100P100 case is much less than that for the H250P250 case.

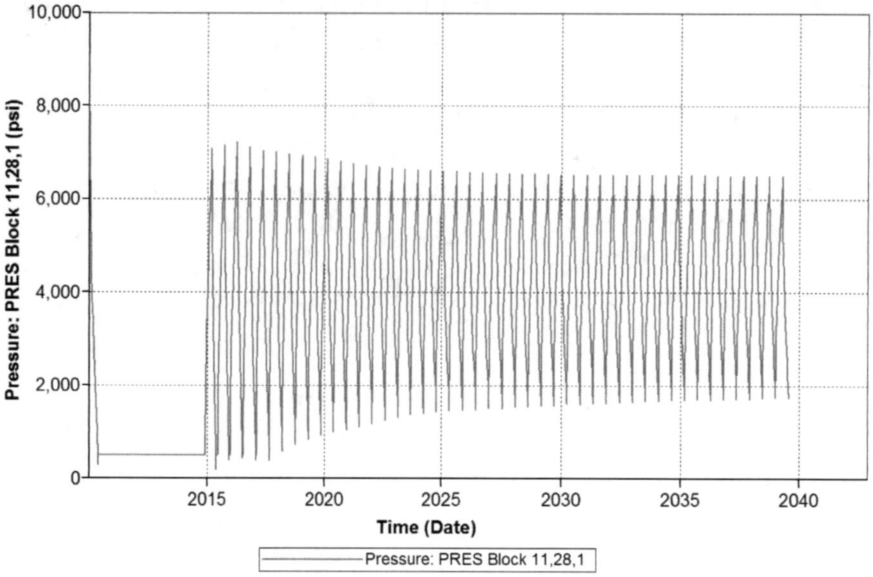

(a) H100P100

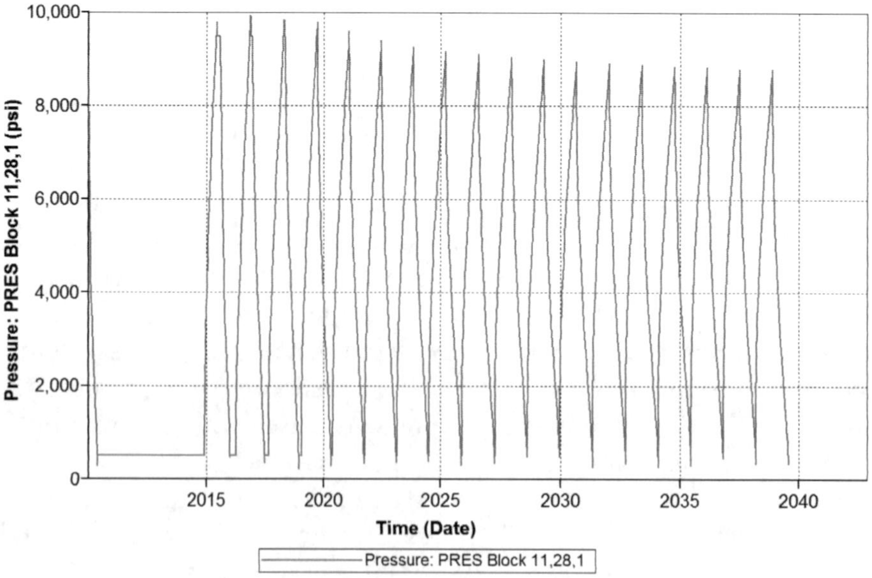

(b) H250P250

Figure 2 Near-Wellbore Block Pressures in the H100P100 and H250P250 cases.

Figure 3 illustrates block pressures at the injection/production well with longer huff and puff times. It is noted that the injection pressure reaches to the set maximum injection pressure

9500 psi and the production pressure drops to the set minimum production pressure 500 psi in the H500P500 and H1500P1500 cases. However, the gas injection rate cannot maintain the set maximum value 300 MSCF/D during the huff time because of the wellbore pressure control. In other words, the gas injection rate cannot be constantly 300 MSCF/D; otherwise, the wellbore pressure will exceed the set maximum value 9500 psi. This explains our hypothesis that the optimum huff time is when the near-wellbore block pressure reaches to the set maximum injection pressure, and the optimum puff time is when the near-wellbore block pressure reaches to the set minimum production pressure.

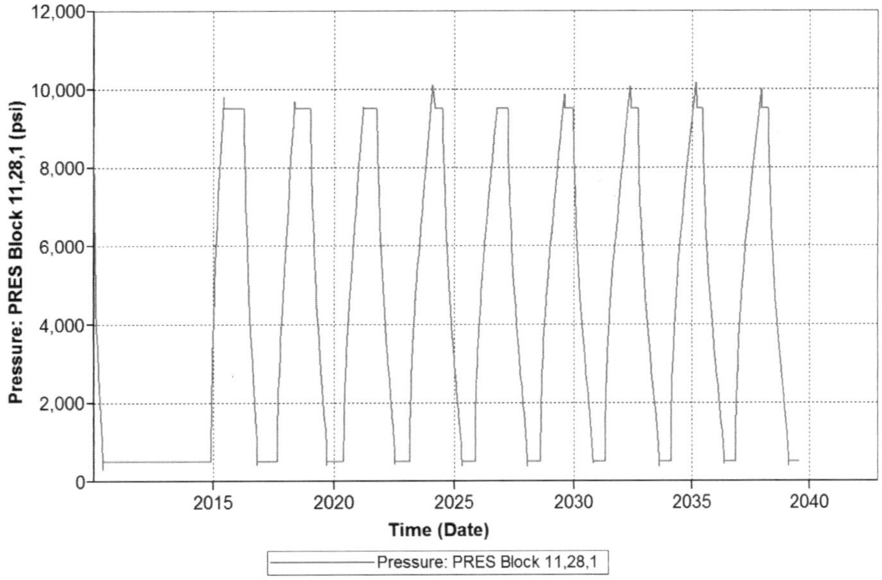

(a) H500P500

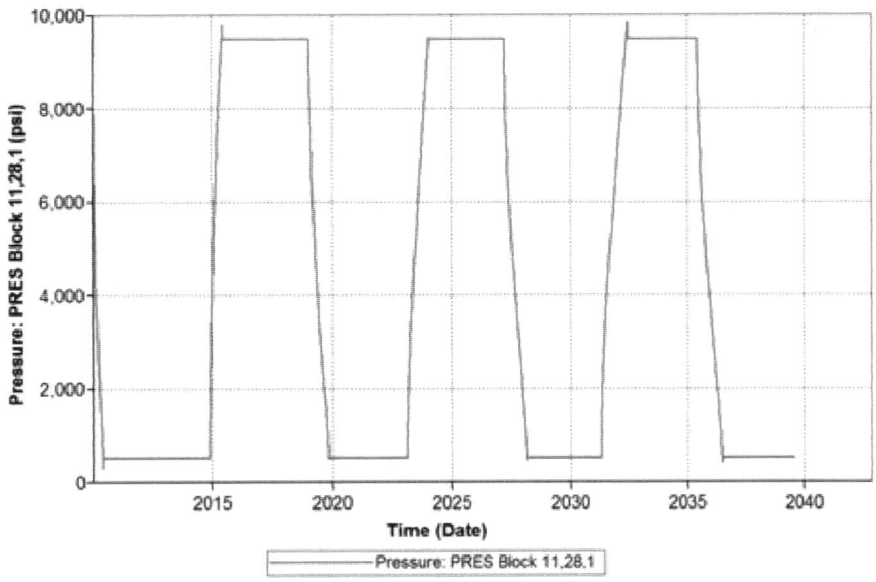

(b) H1500P1500

Figure 3 Near-Wellbore Block Pressures in the H500P500 and H1500P1500 cases.

Moreover, if the hypothesis is valid, the H250P250 case can be further optimized by extending the huff and puff times in the later cycles so as to maintain the set maximum injection pressure through the 25 years of huff-n-puff process. Accordingly, two cases are discussed here. The case H250P250H350P350 is comprised of 4 cycles of H250P250 and 10 cycles of H350P350; whereas the case H250P250H300P300 consists of 6 cycles of H250P250 and 10 cycles of H300P300. The simulation results are shown in Table 1. The oil recovery factors of the case H250P250H350P350 and the case H250P250H300P300 are 34.54% and 34.695%, respectively. The near-wellbore block pressures of these two cases are shown in Figure 4. It is shown that the near-wellbore block pressure reaches to the set maximum injection pressure and the near-wellbore block pressure reaches to the set minimum production pressure in both cases. Therefore, our hypothesis about the optimization of huff-n-puff gas injection in shale gas condensate reservoirs is justified.

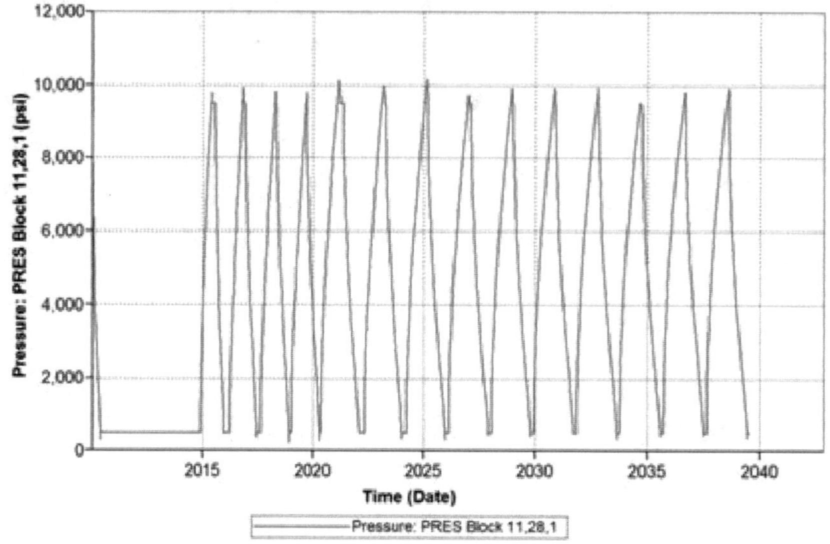

(a) H250P250H350P350

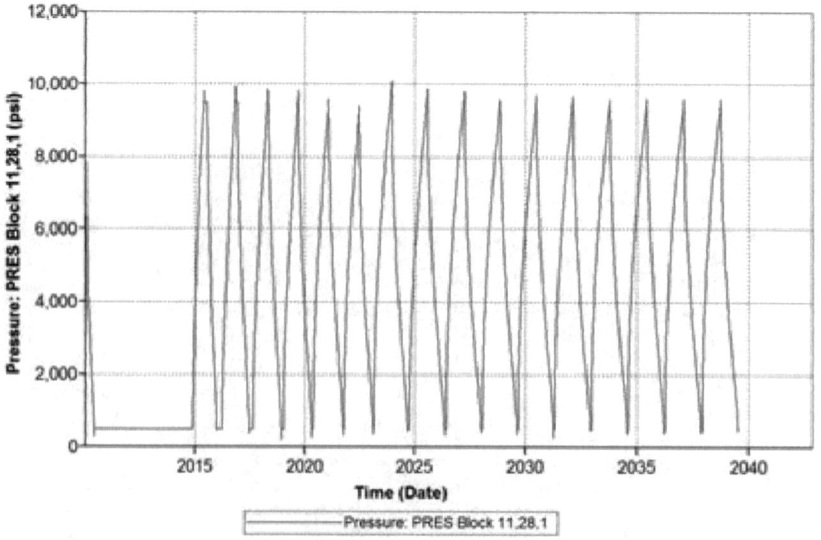

(b) H250P250H300P300

Figure 4 Near-Wellbore Block Pressures in the H500P500 and H1500P1500 cases.

Similarly, different operation conditions are discussed to verify the hypothesis. The set maximum gas injection and production rates are changed to 100 MSCF/D and 600 MSCF/D. The simulation results (Table 3 and Table 4) show that when maximum rate is 100 MSCF/D the optimal case is H1500P1500 (Oil Recovery=28.247%) and when maximum rate is 600 MSCF/D, the optimal case is H125P125 (Oil Recovery=38.259%). The near-wellbore block pressures are investigated and it is found that the near-wellbore block pressure reaches to the set maximum injection pressure and the near-wellbore block pressure reaches to the set minimum production pressure in the optimal cases for different maximum gas rates, as shown in Figure 5. Therefore, it can be concluded that the optimum huff time is when the near-wellbore block pressure reaches to the set maximum injection pressure, and the optimum puff time is when the near-wellbore block pressure reaches to the set minimum production pressure. Furthermore, as the maximum gas injection and production rates increase, the optimum huff and puff times decrease.

Table 3 Effect of Huff and Puff Times on Oil Recovery and Gas Recovery (Qg_{max}=100MSCF/D)

Case	Huff, days	Puff, days	Oil RF, %	Gas RF, %
H100P100	100	100	25.629	32.345
H125P125	125	125	26.025	32.345
H150P150	150	150	26.382	32.345
H180P180	180	180	26.657	32.345
H250P250	250	250	27.032	32.345
H300P300	300	300	27.184	32.345
H500P500	500	500	27.436	32.345
H750P750	750	750	27.895	32.345
H1500P1500	1500	1500	28.247	35.758
H2250P2250	2250	2250	27.995	42.485

Table 4 Effect of Huff and Puff Times on Oil Recovery and Gas Recovery (Qg_{max}=600MSCF/D)

Case	Huff, days	Puff, days	Oil RF, %	Gas RF, %
H100P100	100	100	37.682	20.554
H125P125	125	125	38.259	21.719
H150P150	150	150	38.174	22.725
H180P180	180	180	37.578	23.906
H250P250	250	250	36.058	26.222
H300P300	300	300	35.091	27.681
H500P500	500	500	32.305	32.812
H750P750	750	750	30.171	37.113
H1500P1500	1500	1500	27.178	43.916
H2250P2250	2250	2250	25.725	48.029

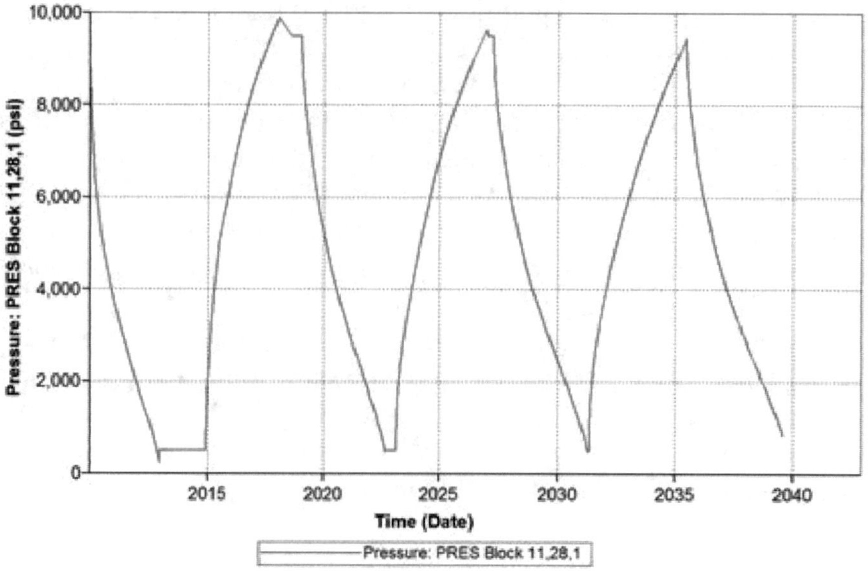

(a) H1500P1500, Qg_{max}=100MSCF/D

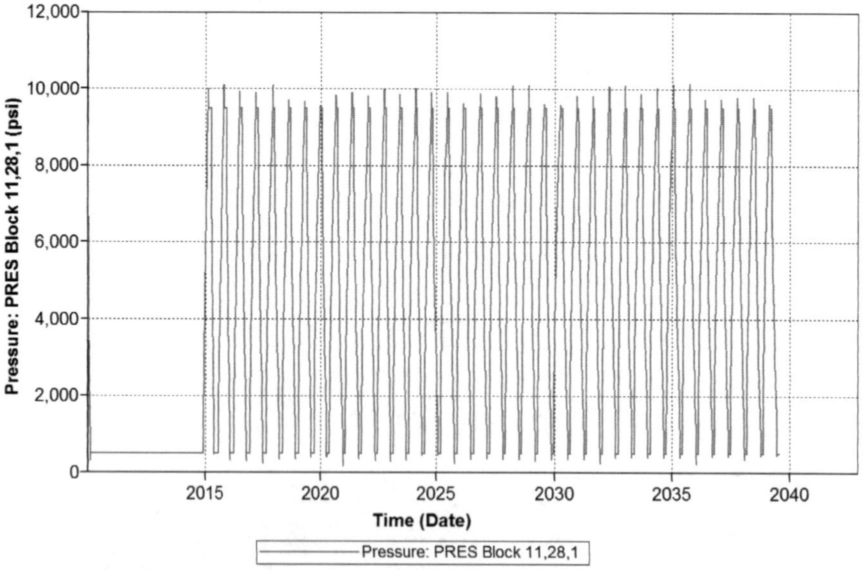

(b) H125P125, Qg_{max}=600MSCF/D

Figure 5 Near-Wellbore Block Pressures in the Optimal Cases when Qg_{max}=100MSCF/D and Qg_{max}=600MSCF/D.

Optimum Soaking Time

Few literatures investigated the soaking time in shale gas condensate reservoirs. However, discusses have been addressed on shale oil reservoirs. Monger and Coma (1988) reported the field tests using the soaking time of 18-52 days, and the results indicated that the effect of soaking time is not significant. Their laboratory tests showed soaking time improved recovery of waterflooding residual oil in the cores and the improved oil recovery was mainly from the subsequent waterflooding period. However, the total time of the waterflooding process with soaking time was longer than the total time without soaking time for those experiments. Gamadi *et al.* (2013) studied the effects of soaking pressure and soaking time in the laboratory. The oil recovery increased with soaking pressure and soaking time. Their results showed that the oil recovery was dependent more on soaking-pressure than soaking-time.

Meng and Sheng's (2015) experimental data showed slightly higher oil recovery with the longer soaking time in shale gas condensate reservoirs. However, the total experimental time with longer soaking time becomes longer. Further study shows that the benefits of soaking may not compensate the decrease in injection and production times due to the time cost in the soaking period.

It may be indicated that soaking time is also beneficial to oil recovery in shale gas condensate reservoirs, whereas the loss in injection and production times due to the time cost in the soaking period may overweight the benefits of soaking. To verify that, the total huff-n-puff process in our model lasts 25 years after 5 years of primary depletion no matter how long is the soaking period, thus the total elapse time is kept constant.

In the base case, the formation permeability is 0.0001 md. By keeping the total elapse time (10800 days) and the huff time (100 days) unchanged, soaking times of 5, 15, 25, 35,45 and 50 days are added in the cases H95S5P100, H85S15P100, H75S25P100, H65S35P100, H55S45P100 and H50S50P100. Table 5 shows that as soaking time increases, the oil recovery factor decreases. The optimum case is H100P100 without soaking, which obtains 32.495% oil recovery. Therefore, soaking time may not be considered in the optimization of huff n puff process. To draw the conclusion, more cases are investigated in this study.

Table 5 Effect of Soaking time (k=0.0001 md)

Shut in time, days	0	5	15	25	35	45	50
Total gas produced (MMSCF)	1490.63	1469.7	1352.3	1229	1103.2	979.6	918.22
Gas injected (MMSCF)	1350	1282.5	1147.5	1012.5	877.5	742.5	675
Net gas produced (MMSCF)	140.63	187.2	204.8	216.5	225.7	237.1	243.22
Gas recovery factor (%)	30.468	40.566	44.373	46.908	48.896	51.367	52.692
Oil produced (MSTB)	37.98	37.592	37.019	35.883	34.634	33.166	32.677
Oil recovery factor (%)	32.495	32.16	31.67	30.70	29.63	28.38	27.96

Figures 6, 7 and 8 show the oil recovery factor at different soaking times as formation permeability varies, which are 0.0001 md, 0.0005 md and 0.001 md respectively. The oil recovery factor decreases as soaking time increases in the 0.0001 md and 0.001 md cases. In the 0.0005 md case, the oil recovery at 5 days soaking time is slightly higher than that at 0 day soaking time. This does not affect the reliability of our conclusion that soaking time should not be added in shale gas condensate reservoirs. After all, a time gap of a few days exists between injection and production in experiment and field projects because of operation requirement.

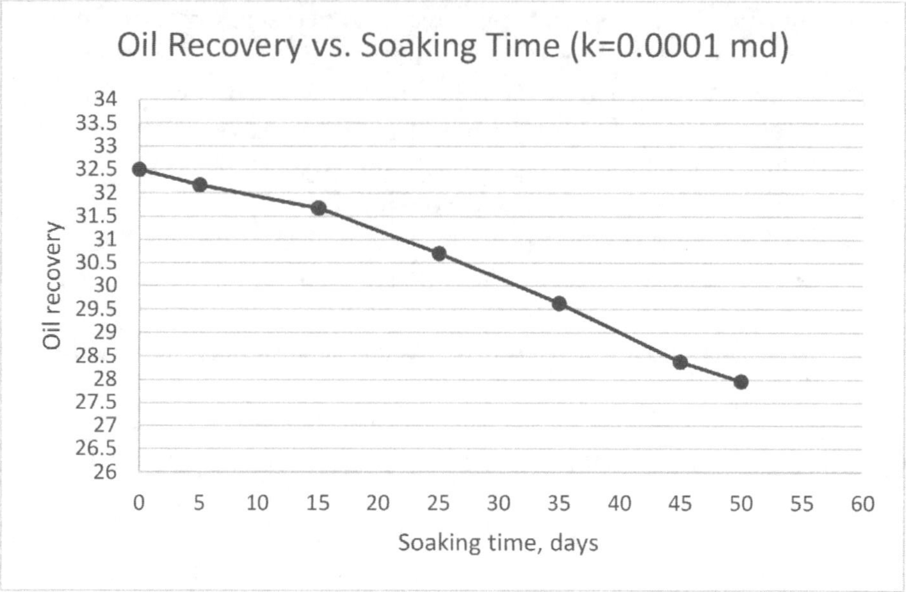

Figure 6 Oil Recovery at Different Soaking Times (k=0.0001 md).

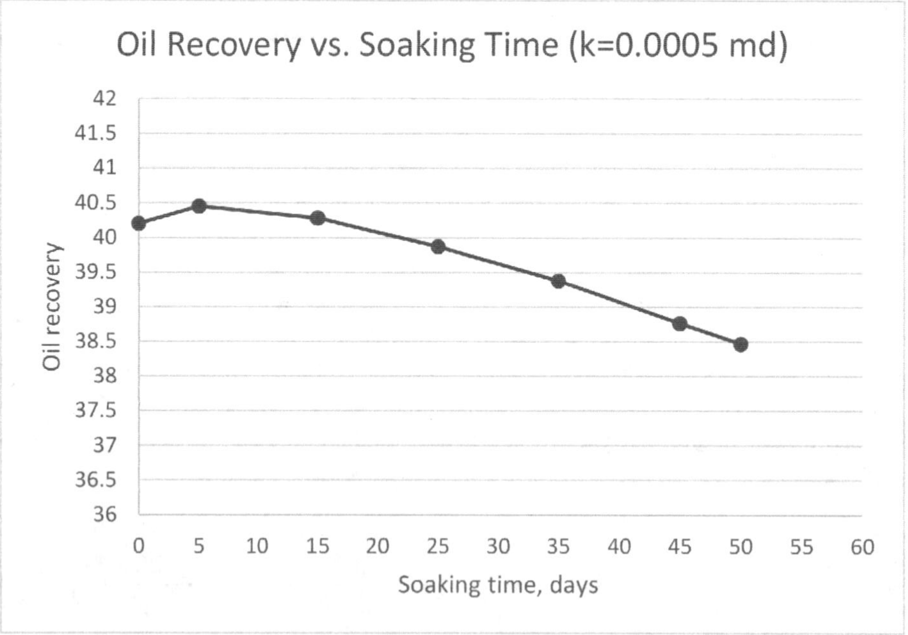

Figure 7 Oil Recovery at Different Soaking Times (k=0.0005 md).

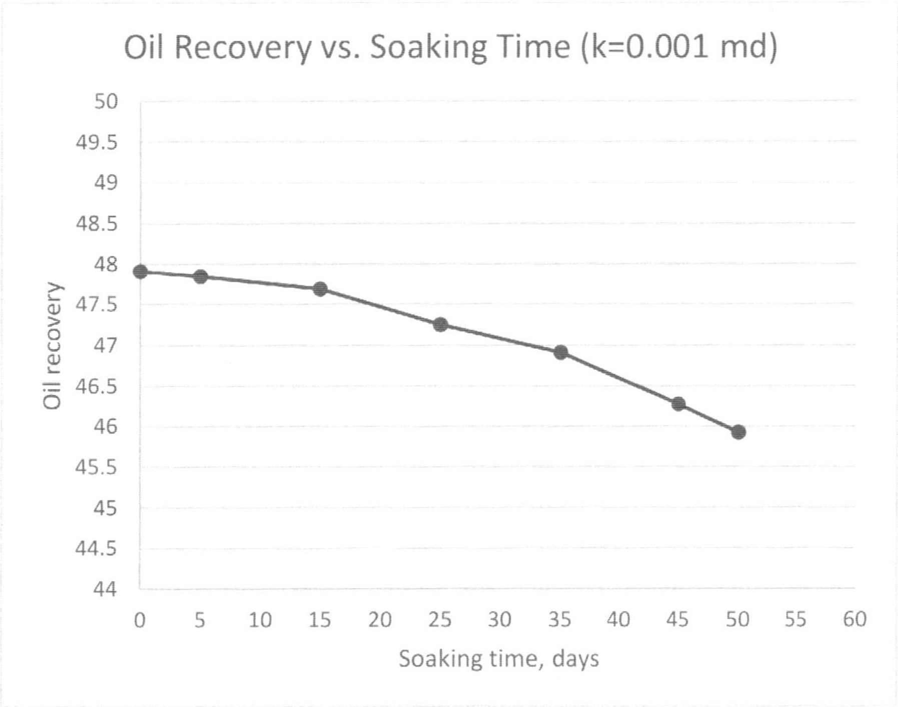

Figure 8 Oil Recovery at Different Soaking Times (k=0.001 md)

Conclusions

Based on this simulation study, the following conclusions may be drawn.

1. The maximum injection rate and the maximum injection pressure during the huff period, and the maximum gas and oil production rates and the minimum production pressure during the puff period are determined by the facility capacities, reservoir conditions, and operation constraints. The optimum huff time is so long that the near-wellbore pressure reaches the set maximum injection pressure during the huff period; and the optimum puff time is the time required for the near-wellbore pressure to reach the set minimum production pressure.
2. As the maximum gas injection and production rates increase, the optimum huff and puff times decrease.
3. The benefits of soaking may not compensate the loss in injection and production due to the time lost in the soaking period. Therefore, soaking may not be necessary during the huff-n-puff gas injection in shale oil reservoirs.

References

1. Orangi, A., Nagarajan, N. R., Honarpour, M. M., & Rosenzweig, J. J. (2011, January 1), "Unconventional Shale Oil and Gas-Condensate Reservoir Production, Impact of Rock, Fluid, and Hydraulic Fractures," Society of Petroleum Engineers, doi:10.2118/140536-MS.
2. James J. Sheng (2015), "Increase Liquid Oil Production by Huff-n-Puff of Produced Gas in Shale Gas Condensate Reservoirs," Journal of Unconventional Oil and Gas Resources 11 (2015) 19-26.
3. Kurtoglu, B., Sorensen, J. A., Braunberger, J., Smith, S., & Kazemi, H. (2013, August 12), "Geologic Characterization of a Bakken Reservoir for Potential CO2 EOR," Society of Petroleum Engineers, doi:10.1190/URTEC2013-186.

4. Shoaib, S., & Hoffman, B. T. (2009, January 1), "CO2 Flooding the Elm Coulee Field," Society of Petroleum Engineers, doi:10.2118/123176-MS.
5. Wang, X., Luo, P., Er, V., & Huang, S.-S. S. (2010, January 1), "Assessment of CO2 Flooding Potential for Bakken Formation, Saskatchewan," Society of Petroleum Engineers, doi:10.2118/137728-MS.
6. Monger, T. G., & Coma, J. M. (1988, November 1), "A Laboratory and Field Evaluation of the CO2 Huff 'n' Puff Process for Light-Oil Recovery," Society of Petroleum Engineers, doi:10.2118/15501-PA.
7. Gamadi, T. D., Sheng, J. J., & Soliman, M. Y. (2013, September 30), "An Experimental Study of Cyclic Gas Injection to Improve Shale Oil Recovery," Society of Petroleum Engineers, doi:10.2118/166334-MS.
8. Meng, X., Yu, Y., Sheng, J. J., Watson, M., & Mody, F. (2015, August 4), "An Experimental Study on Huff-n-Puff Gas Injection to Enhance Condensate Recovery in Shale Gas Reservoirs," Society of Petroleum Engineers, doi:10.15530/urtec-2015-2153322.

434257 Gas Permeability in Natural Media

Abstract:

With a sustained interest in exploration for shale gas, and condensate recovery from ultra-tight reservoirs, for purposes of understanding production decline, the mechanism of transport within the reservoir matrix is of fundamental importance. In our laboratory, a permeameter to measure rock permeability down to a hundred nanoDarcy (0.001 nm^2) has been built and tested rigorously. In this device, the experimental protocol calls for a perturbation from an equilibrated pressure. The resulting decay in the imposed perturbation to a final equilibrium value is used to infer the permeability of the sample. The details of the methodology is available elsewhere (paper under review in AIChEJ.)

In this paper, we study effects of various factors such as stress and fluid pressure on the permeability of the sample. We show that two distinct contributions of fluid pressure act in opposing directions. The first is pore-wall slip, which decreases with increasing pressure, and thus reduces permeability with pressure. The second is effective stress which decreases with increasing pressure, and therefore tends to favor permeability increase with pressure. Since the dependence of permeability on stress is path dependent, an experimental protocol to unambiguously determine this variation is quite to difficult to construct, and we discuss possible methods.

For characterization purposes, we also show that some tight rock samples have permeability variation with respect to molecular species. We conjecture that this variation is caused by adsorption. The mechanism of permeability alteration due to adsorption is debatable, and procedures to separate the contribution as that due to effective pore size are discussed.

Enhancing the Predictive Capability of the Pore-Scale Multiphase Fluid Transport in Unconventional Reservoirs: A Molecular Dynamics Simulation Approach

Gorakh Pawar, Ilija Miskovic

Subsurface Transport Phenomena Lab,
University of Utah, Salt Lake City, Utah, USA-84112

AIChE 2015, November 10, 2015
Salt Lake City, Utah, United States

STPL

*Email: ilija.miskovic@utah.edu
gorakh.pawar@utah.edu

Overview

- Background
- Development of 3D kerogen matrix models with a submerged nanometer and sub-nanometer sized pores
- Molecular dynamics simulation of multiphase flow in kerogen pores
- Results and discussion
- Conclusion

2

Research background

- **The key challenges** in unconventional reservoirs are
  - ❑ Low recovery rates
  - ❑ Most of the hydrocarbons are trapped in adsorbed and absorbed form (in kerogen matrix)
  - ❑ Highly complex and poorly connected pore network structure
  - ❑ Limitation of the continuum theory describing the flow transport
- A **framework** is necessary that can capture the **fine details** of the multiphase transport in kerogen nanopores, and translate this expertise to design and development of new methods that will enhance the **hydrocarbon recovery rates** from unconventional reservoirs

3

Identification of the ROI from pore network in kerogen matrix

- Figure 1 shows the reconstructed FIB/SEM images of the **representative shale sample**

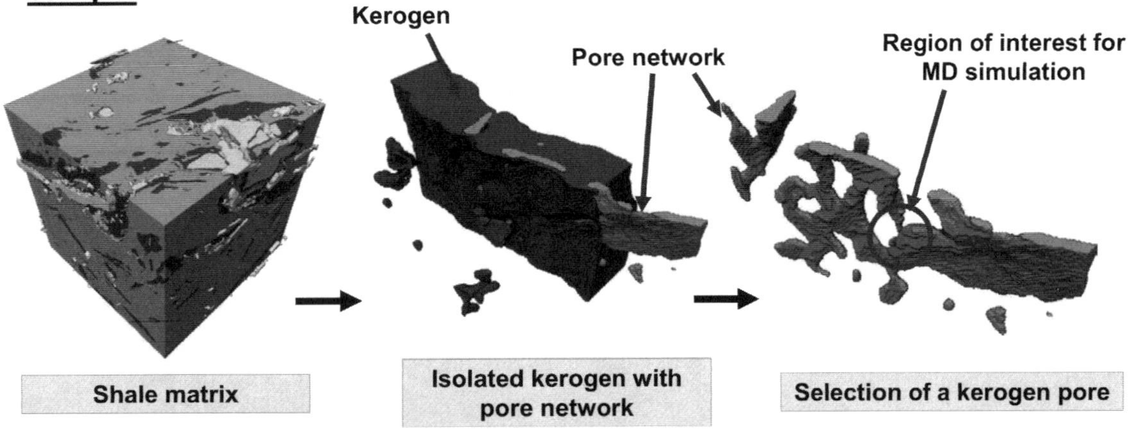

Figure 1: Identification of the region of interest (ROI) from pore network in kerogen matrix

4

Challenges developing kerogen model with submerged pores

- Replication of the **thermo-geo-chemical-mechanical transformation of organic matter** that happened over millions of years (geological timescale) and resulted in a kerogen structure with submerged nanopores
- Mostly **carbon-based models**[1,2] used to represent the kerogen
- More than one **factors affecting the pore-scale fluid flow** which includes the
  - ❑ Thermal maturity of kerogen and total organic content
  - ❑ Pore skeleton
  - ❑ Pore composition
  - ❑ Pore surface anisotropy
  - ❑ Pore injection pressure and
  - ❑ In-situ reservoir conditions

5

Molecular models of kerogen with submerged pores

- Figure 2 depicts the procedure to develop the molecular model of kerogen with submerged nanopores

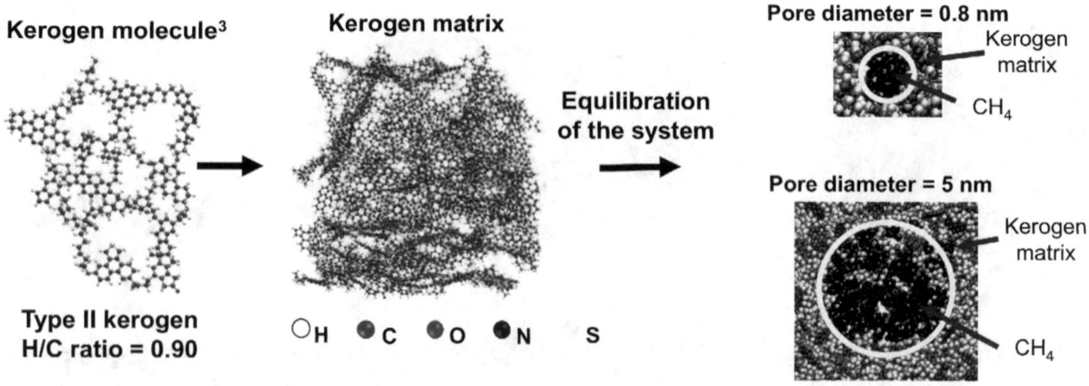

Figure 2: The development molecular model of kerogen matrix with submerged nanopore

6

Molecular dynamics simulation framework

Define pore configuration

- Pore size
- Constitutive atom types and a total number pf atoms
- Fundamental atomic properties (charge, mass, interatomic potential)

Define/Calibrate MD simulation parameters

- Initial state of a system (position, velocity)
- Time step (Δt) and total simulation time
- Thermostat properties

Data logging and analysis

- Obtain atom trajectories at different time steps
- Evaluate qualitative and quantitative system properties
- Visualization of results

Figure 3: Molecular dynamics simulation stages

Total potential energy consists of rotational, translational, and vibrational motion. Further, the van der Waals forces and electrostatic interactions also contribute to the total energy

$$F_i = m_i \frac{d^2 r_i}{dt^2} \rightarrow F_i = \nabla(U(r_i)) \rightarrow U = E_b + E_{nb}$$

$$E_B = \frac{1}{2} K_e (R - R_e)^2$$

$$E_C = C \frac{q_i q_j}{\epsilon r}$$

$$E_A = \frac{1}{2} K_\theta (\theta - \theta_e)^2$$

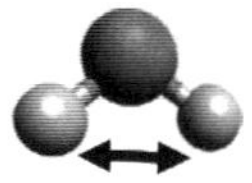

$$E_{vdW} = 4\varepsilon \left[\left(\frac{\sigma}{r}\right)^{12} - \left(\frac{\sigma}{r}\right)^{6} \right]$$

$$E_T = \frac{1}{2} V\{1 - cos[n(\emptyset - \emptyset_e)]\}$$

7

Interatomic potential and simulation details

DRIEIDING force field

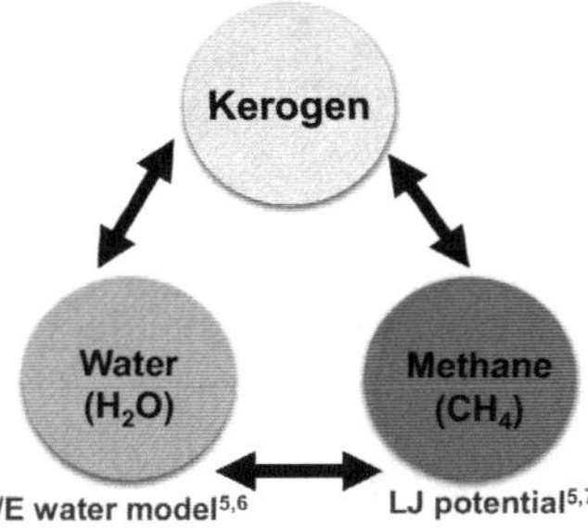

SPC/E water model[5,6] LJ potential[5,7]

Figure 4: The force fields used in the MD simulation

- **A general-purpose DRIEIDNG force field[4]** used to where there is no or very little experimental data available (applicable to kerogen modeling)

Table 1: Number of molecules used in the MD simulation

Parameter	Pore diameter			
	0.8 nm	**1 nm**	**3 nm**	**5 nm**
Number of H_2O/CH_4 molecules	25	40	100	250
Number of kerogen units	60	60	60	60

Table 2: Force applied to water molecules and corresponding pore pressure

Force (Kcal/mol-Å)	Pore pressure (psi)			
	0.8 nm	**1 nm**	**3 nm**	**5 nm**
0.05	25,070	25,676	7,132	6,419
0.1	50,140	51,352	14,264	12,838

8

Additional simulation details

- LAMMPS[8] used to perform MD simulations
- NVT ensemble used with system temperature of 340 K
- Adaptive time stepping is used, and output data logged every 2000 time steps
- Dynamic properties of the injected (water) and the recovered fluid (methane) are evaluated for various pore diameters and compared
  - ❑ Mean square displacements of methane and water
  - ❑ Effective diffusion coefficients of methane and water
  - ❑ Distribution of methane molecules within the pore

Results : Fluid transport in 0.8 nm kerogen pore

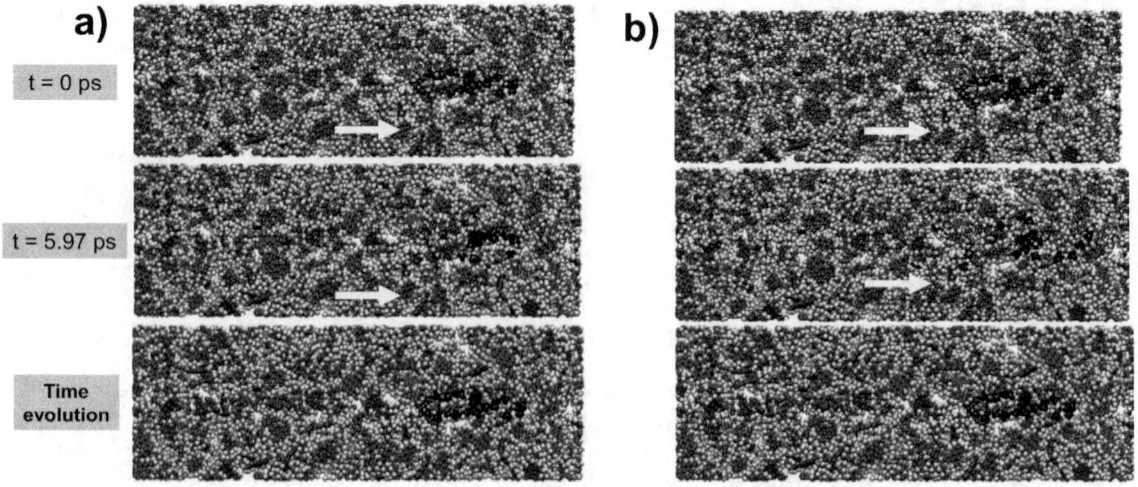

Figure 5: Fluid flow in 0.8 nm kerogen pore at force of a) 0 Kcal/mol-Å b) 0.1 Kcal/mol-Å

Fluid transport in 5 nm kerogen pore

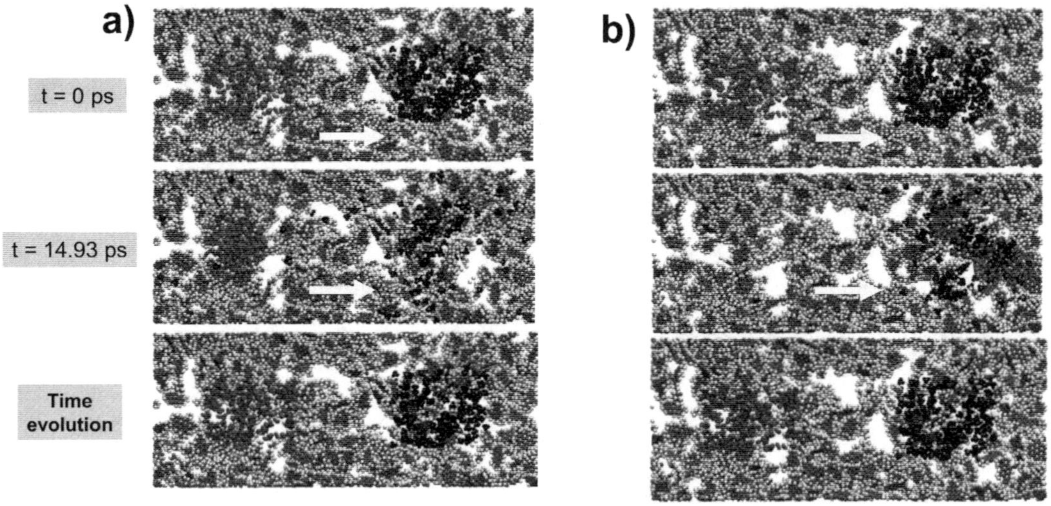

Figure 6: Fluid flow in 5 nm kerogen pore at force of a) 0 Kcal/mol-Å b) 0.1 Kcal/mol-Å

Mean square displacement of water and methane

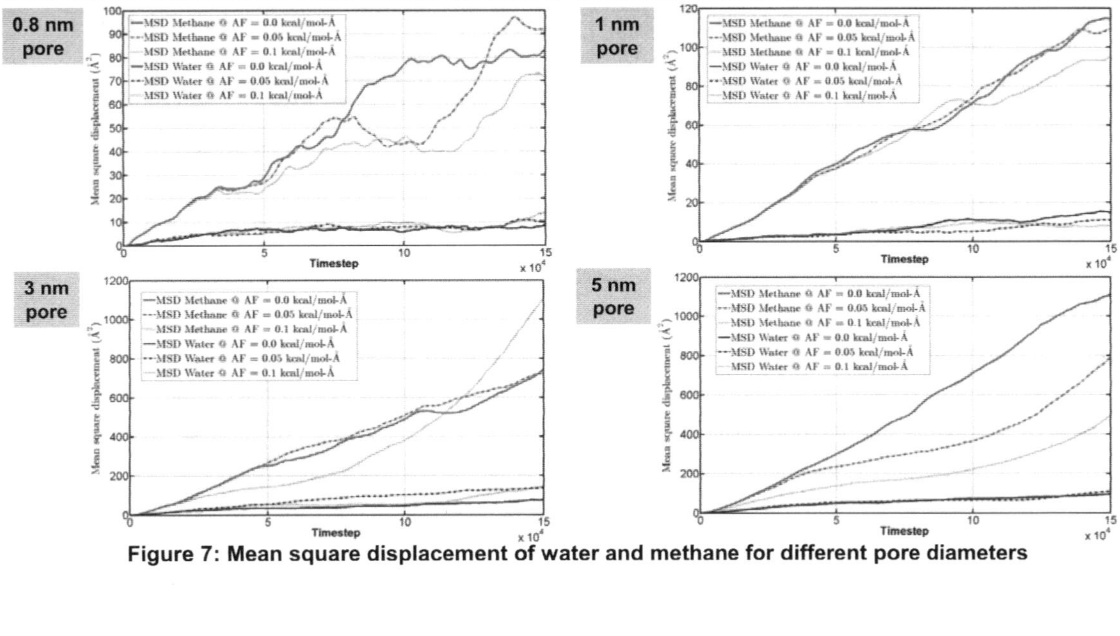

Figure 7: Mean square displacement of water and methane for different pore diameters

Effective diffusion coefficients

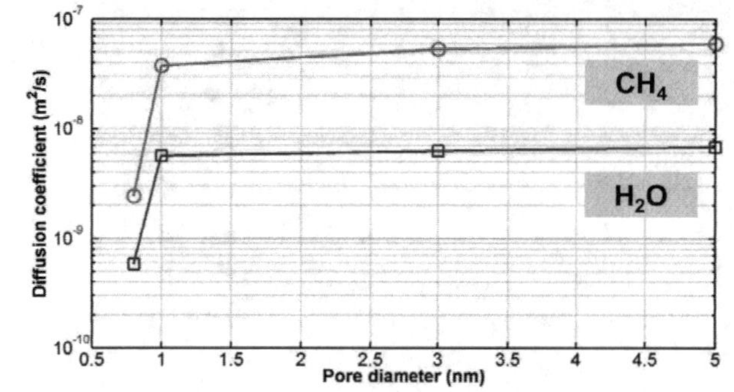

Figure 8: Diffusion coefficient variation of methane and water

Table 2: Diffusion coefficient comparison

Pore Diameter (nm)	D (CH_4) m^2/s	D (H_2O) m^2/s
0.8	2.43E-09	5.85E-10
1	3.77E-08	5.67E-09
3	5.33E-08	6.29E-09
5	5.94E-08	6.81E-09

- Effective diffusion coefficient of **CH_4 is higher than H_2O**
- Effective diffusion coefficient of both CH_4 and H_2O **increases with increase in the pore diameter but the trend is not linear**

13

Methane number density in 0.8 nm kerogen pore

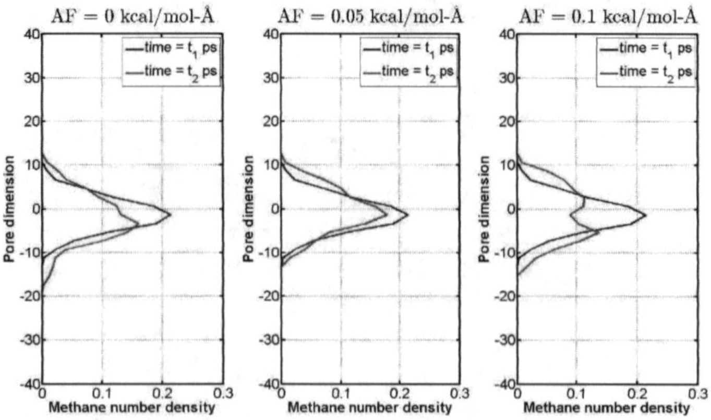

Figure 9: Methane number density in 0.8 nm pore

For applied force of 0 Kcal/mol-Å

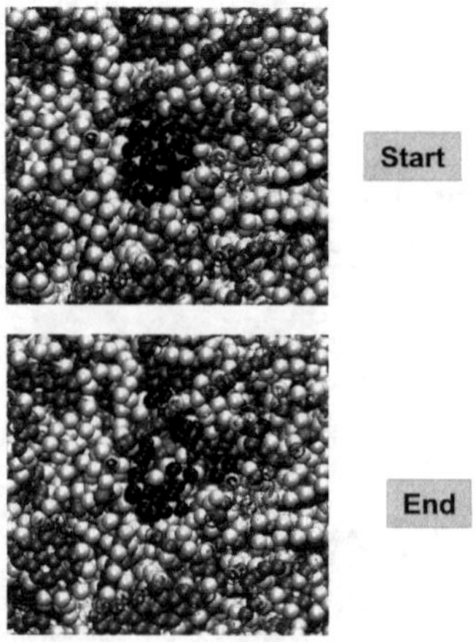

14

Methane number density in 5 nm kerogen pore

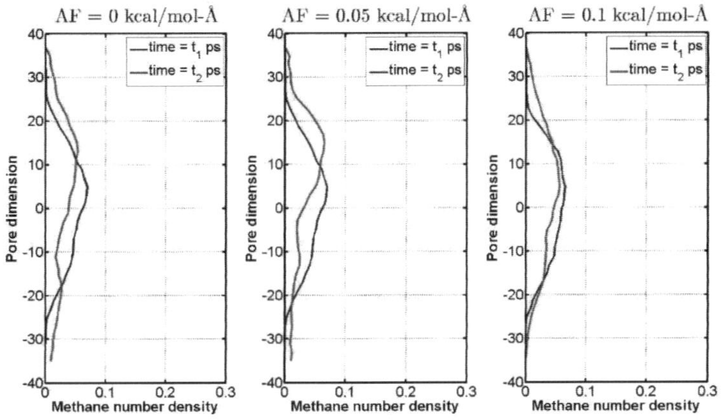

Figure 10: Methane number density in 5 nm pore

For applied force of 0 Kcal/mol-Å

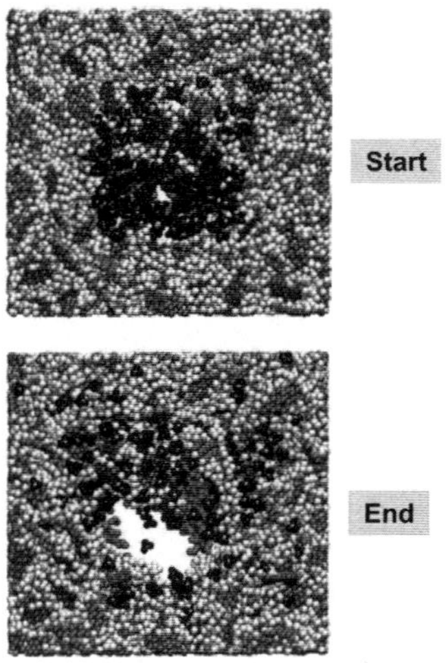

Methane number density in 1 nm and 3 nm pores

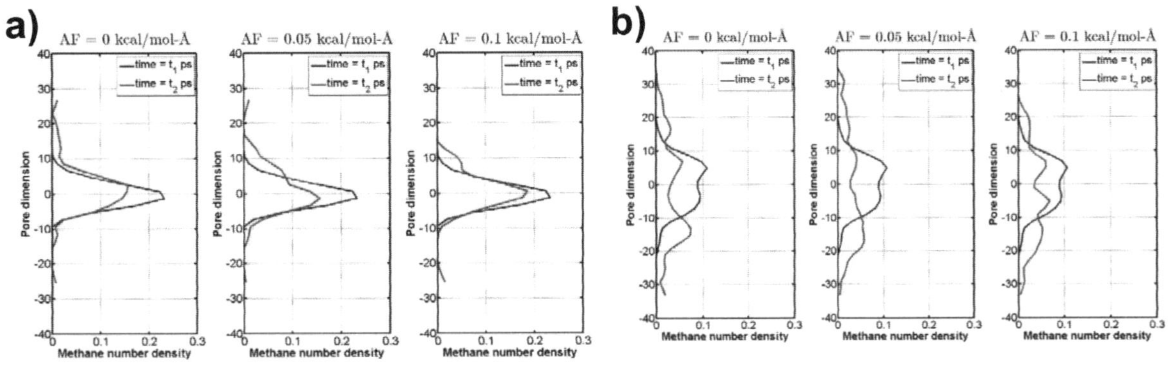

Figure 11: Methane number density in a) 1 nm and b) 3 nm kerogen pores

Conclusion

- Developing **kerogen molecular models with submerged nanopores** is a complex process
- MD simulations **enable access to length scales** (sub-nanometer and nanometer sized pores) which is challenging otherwise
- Gas exists in adsorbed as well as free form – free gas is primarily recovered
- **Higher injection and/or differential pore pressures** are necessary to overcome a dominant fluid-pore surface interactions and **to enhance the mobility of fluids** in nanometer and sub-nanometer sized kerogen pores
- Evidence of water and methane adsorption on the pore surface

17

Acknowledgement

We would like to thank Dr. Paul Meakin, Temple University and Dr. Hai Huang, Idaho National Laboratory for insightful discussion and comments. Also, we would like to thank Idaho National Laboratory for providing the access to the Falcon supercomputer

18

References

1. Jin, Zhehui, and Abbas Firoozabadi. "Flow of methane in shale nanopores at low and high pressure by molecular dynamics simulations." The Journal of chemical physics 143.10 (2015): 104315.
2. Hu, Y. et al., 2013. A pore scale study describing the dynamics of slick water distribution in shale gas formations following hydraulic fracturing, SPE Unconventional Resources Conference-USA. Society of Petroleum Engineers.
3. Ungerer, P.; Collell, J.; Yiannourakou, M., Molecular modeling of the volumetric and thermodynamic properties of kerogen: influence of organic type and maturity. Energy & Fuels 2014, 29 (1), 91-105
4. Mayo, S. L.; Olafson, B. D.; Goddard, W. A., DREIDING: a generic force field for molecular simulations. Journal of Physical Chemistry 1990, 94 (26), 8897-8909
5. S.K. Reed, R.E. Westacott, The interface between water and a hydrophobic gas, Physical Chemistry Chemical Physics, 10 (2008) 4614-4622.
6. J. Zielkiewicz, Structural properties of water: Comparison of the SPC, SPCE, TIP4P, and TIP5P models of water, The Journal of chemical physics, 123 (2005)
7. H. Docherty, A. Galindo, C. Vega, E. Sanz, A potential model for methane in water describing correctly the solubility of the gas and the properties of the methane hydrate, The Journal of chemical physics, 125 (2006) 074510
8. S. Plimpton, Fast Parallel Algorithms for Short-Range Molecular Dynamics, J Comp Phys, 117, 1-19 (1995)

19

THANK YOU

425791 A Combined Asphaltene Study Relating Kinetic Aggregation and Deposition Parameters to Thermodynamic Driving Forces

Abstract:

This study discusses experimental results of asphaltene aggregation and deposition using various n-alkanes as precipitants to destabilize asphaltenes. The amount of asphaltenes destabilized with different precipitants and at various concentrations was obtained, and the collision efficiency for asphaltene aggregation was calculated using a geometric population balance. The results revealed that for a fixed volume fraction of precipitant, the asphaltene-asphaltene collision efficiency decreased with higher carbon number precipitants. Furthermore, decreasing the precipitant concentration resulted in lower collision efficiencies, in agreement with previous studies. A correlation was developed to relate between the collision efficiency and a thermodynamic driving force of the differences in asphaltene and surrounding liquid solubility parameters with various precipitants and precipitant concentrations. In addition, the tendency for asphaltenes to deposit was measured using capillary flow experiments with varied precipitants and precipitant concentrations. The results revealed that the time for asphaltene deposits to be detected can be estimated using a similar solubility parameter correlation. The results also demonstrated that the deposition behavior of asphaltenes can be normalized by considering the concentration of precipitated asphaltenes flowing through the apparatus. Scanning electron microscopy (SEM) images of the asphaltene deposits were obtained and used to assess the mixing of oil and precipitant. This study unifies two independent asphaltene processes under a single thermodynamic framework.

426554 The Impact of Shear and Solid Stabilizers on Cyclopentane Hydrate Formation in Emulsions

Abstract:

Hydrates are of great importance in the oil and gas industry due to their ability to plug pipelines, thereby leading to significant costs involved with managing and mitigating associated flow assurance issues. Along with naturally occurring surfactants, a wide variety of solid particles are typically present in crude oil. The impact of these solids on hydrate formation must be properly understood in order to develop effective flow assurance management strategies. The lack of fundamental understanding of hydrate formation mechanisms in emulsions has motivated the current work focused on hydrate forming emulsions stabilized using both conventional surfactants and solid particles. The objective of this work is to provide insight into hydrate formation in emulsions by quantifying droplet and hydrate particle size using optical microscopy. For this work, an Olympus BX53 polarized optical microscope with shear cell and temperature control stage (-50°C to 450°C) is used to quantify droplet size distributions and hydrate particle morphology. Cyclopentane hydrate formation is quantified in model oil systems stabilized using both solid particles and conventional surfactants. The concentrations of the dispersed phase and the surfactant are varied to evaluate their effect on hydrate formation and emulsion behavior.

422926 Measurement and Modeling of Crude Oil Density at Temperatures to 250° C and Pressures to 275 Mpa

Abstract:

Fundamental properties of crude oils such as density are important in the estimation of the amount of recoverable petroleum in an oil reserve. In addition, the isothermal compressibility, which is a density-derived property, is needed for estimating the fraction of the total hydrocarbons present that can be produced in the primary recovery period.

We have investigated the performance of several equation of state (EoS) models for predicting the high-temperature, high-pressure (HTHP) volumetric properties of hydrocarbons and their mixtures. This HTHP designation, which encompasses, temperatures to 250 ˚C and pressures to 275 MPa, covers the conditions encountered on the platform as well as ultra-deep formation conditions such as those that exist beneath the deep sea floor in the Gulf of Mexico. As such, the density data for two dead crude oil samples from the Gulf of Mexico have been measured from ambient to HTHP conditions at temperatures to 250 ˚C and pressures to 275 MPa. The density measurement for simulated live crude oil samples is in progress, where the simulated live oil is obtained by adding gases to the two aforementioned dead oil samples.

A variable-volume view cell was used in the measurement of the crude oil density data. The view cell has previously been successfully used to measure the density data for various pure component and binary hydrocarbon mixtures at temperatures to 250 °C and pressures to 275 MPa. From the experimental density data for the crude oil samples, the isothermal compressibility values for the oils are then calculated as a function of pressure. The perturbed-chain statistical associating fluid theory (PC-SAFT) EoS has been used to successfully predict the density and isothermal compressibility of the crude oils. The performance of the PC-SAFT EoS is studied using the sets of original PC-SAFT parameters and the sets of parameters obtained from the fit of the PC-SAFT EoS to HTHP pure component density data. To obtain PC-SAFT parameters for the crude oil pseudocomponents, correlations were used wherein the pure component parameter values are defined as a function of molecular weight and aromaticity. Preliminary results show that the predictions using the PC-SAFT EoS with the HTHP parameters are in good agreement with the isothermal compressibility data to within 7%. This number represents a significant improvement over the 25% agreement obtained when using the original set of PC-SAFT parameters. In summary, the PC-SAFT EoS predictions, particularly those for the isothermal compressibility, show great promise over those predicted by traditional cubic EoSs.

430044 Solvent Effects on the Structure of Petroleum Asphaltenes

Abstract:

The asphaltene nanoaggregate structure in liquid environments has been recently shown to differ from that of solid asphaltenes [1]. The nature of asphaltene-asphaltene interactions is manifested in their associated structure, which was shown to change in various solvents, and prompts further study. Neutron diffraction provides a quantitative experimental description of the influence of the surrounding liquid matrix on structure of asphaltene nanoaggregates in mixtures of solvents and anti-solvents. In addition, neutron diffraction allows for isotopic substitution of the solvent and investigation of the interaction of hydrogen atoms, which are virtually transparent to X-rays. This talk will present first of their kind neutron diffraction results that were generated for samples containing 5 vol. % asphaltenes dissolved in various deuterated solvents and anti-solvents on the NOMAD instrument at the Spallation Neutron Source at Oak Ridge National Lab. Aromatic, polar and halogenated compounds were used as the solvents, while n-heptane served as the anti-solvent. Results were instrument and solvent background corrected to isolate the scattering of the asphaltene molecules and their aggregated structures. This investigation differs from previous small-angle scattering studies by examining the molecular spacing between asphaltenes on the angstrom scale as opposed to viewing larger (e.g., nanometer) associations with small-angle scattering. The separation distance of each diffraction peak was calculated and compared to previous X-ray scattering results [1]. In the neutron diffraction results, the peaks corresponding to the organization of alkyl side chains and aromatic stacking were observed in agreement with previous X-ray results. A new correlation peak was identified at a separation distance of approximately 2 angstroms that corresponds to the presence of hydrogen bonding. The novel results generated with neutron diffraction reinforce the need to consider the local solvent matrix on the structure and interactions of asphaltenes.

[1] Hoepfner, M. P. and Fogler, H. S., "Multiscale Scattering Investigations of Asphaltene Cluster Breakup, Nanoaggregate Dissociation, and Molecular Ordering," Langmuir, 29 (49), 2013.

426182 Modeling of Asphaltene Deposition in Pipelines: Accounting for a Chemical Inhibitor Effect

Abstract:

Asphaltene deposition occurs in production flow lines when pressure drops below the asphaltene onset precipitation pressure, which is determined by oil chemical composition. The deposit layer can, eventually, become thick enough to significantly reduce production rate. Asphaltene deposition can be prevented or reduced by chemical inhibitors. To decide whether using an inhibitor is necessary or not, a reliable deposition model is required. The deposition process includes a number of phenomena, such as precipitation of asphaltene nanoparticles from a fluid, growth of agglomerates, their transport to the wall and agglomerate-wall interaction. The deposit layer formed on the wall can be also partially removed with a fluid flow (the shear removal effect). A deposition model that we developed assumes that both particle agglomeration and particle transport to the wall are governed by Brownian motion. Agglomerate growth is modeled by Population Balance technique. Sticking of a particle to the wall is assumed to be a probabilistic process. Shear removal is described by a purely empirical sub-model. To identify model coefficients for a specific oil, experiments are conducted utilizing a laboratory Taylor-Couette device in which the inner cylinder rotates whereas and the outer cylinder remains immobile. Deposition occurs on the outer cylinder wall. The model developed allows relatively accurate matching of modeling results to experimental data obtained in the laboratory Taylor-Couette device. Since the model is derived from a physical basis, accounting for the effect of asphaltene inhibitors does not require the introduction of an additional model or terms. Inhibitor effects are accounted for through tuning of the existing model coefficients. A model of asphaltene deposition in pipe flow, based on the same principles as that for a Taylor-Couette device, is used for forecasting deposit layer formation on a field scale.

431017 Enthalpy of Solution Differences Between Chemically Separated Asphaltenes and Physically Separated Asphaltene-Rich Materials in Athabasca Bitumen

Abstract:

Asphaltenes, defined as a fraction of crude oils that can be filtered from admixtures with n-alkanes[1] are a surrogate for nanoaggregated materials present[2] with which they share many but not all properties. These nanoaggregated materials, however defined, comprise up to 20 wt % of heavy oils. Once destabilized, they may separate as a solid or viscous liquid phase and stick to surfaces in reservoirs, pipelines and refining process equipment. Deposition of asphaltene-rich materials can cause wellbore or pipeline plugging, degradation of process performance and process failure. During bitumen extraction, asphaltene-rich materials absorb at oil/water interfaces and stabilize oil in water (O/W) or water in oil (W/O) emulsions.[3] Stabilization of water in oil emulsions by asphaltene-rich materials results in inefficient separation of water from the oil phase. Since water drops in these emulsions contain dissolved salts, inefficient separation can cause equipment corrosion in downstream processes. Incomplete removal of oil drops from the water phase reduces the recovery of bitumen and causes difficulties during wastewater treatment.

Many studies, both experimental and theoretical, purport to address the conditions under which asphaltenes or asphaltenes-rich materials deposit on substrates and the extent of deposition. Frequently, the formation of asphaltene phases, as a dispersion, and their deposition on surfaces is not discriminated and correlations identifying conditions where deposition occurs remain unreliable when applied in the field.[4–8] This uncertainty contributes to the technical and hence the economic risk associated with process design and process operation. Further, recent studies have shown that the properties, compositions and phase behaviors of asphaltenes separated from crude oils upon addition of n-pentane, namely C_5 asphaltenes, or n-heptane, C_7 asphaltenes, on which correlations are largely based, are not representative of the asphaltene-rich materials that aggregate naturally in crude oils.[5,9]

In this study the enthalpies of solution of chemically separated C_5 asphaltenes, and physically separated asphaltene-rich nanoaggregated materials are compared in toluene, heptane, and de-ionized water using a precision solution calorimeter. The repeatability and reliability of data obtained with this equipment was validated in a recent prior study.[10] Physical separation of asphaltene-rich nanoaggregated materials from Athabasca bitumen was conducted using nanofiltration with commercially available ceramic membranes at 473 K, with a moderate differential pressure and without solvent addition.[2] The solution behaviors of chemically separated C_5 asphaltenes, and physically separated asphaltene-rich materials are shown to be both quantitatively and qualitatively different. This additional difference between the behaviors of chemically separated C_5 asphaltenes and asphaltene-rich nanoaggregated materials in crude oil underscores the need to work directly with materials in oils that are aggregated or that form aggregates, and not with operationally defined fractions with poorly correlated properties. By doing so, there is potential to develop generalizable mitigation strategies and reliable predictive tools for "asphaltene deposition" in field applications.

(1) Buenrostro-Gonzalez, E.; Lira-Galeana, C.; Gil-Villegas, A.; Wu, J. Asphaltene Precipitation in Crude Oils: Theory and Experiments. AIChE J. 2004, 50, 2552–2570.

(2) Zhao, B.; Shaw, J. M. Composition and Size Distribution of Coherent Nanostructures in Athabasca Bitumen and Maya Crude Oil. Energy & Fuels 2007, 21, 2795–2804.

(3) Yang, X.; Hamza, H.; Czarnecki, J. Investigation of Subfractions of Athabasca Asphaltenes and Their Role in Emulsion Stability. Energy & Fuels 2004, 18, 770–777.

(4) Likhatsky, V. V.; Syunyaev, R. Z. New Colloidal Stability Index for Crude Oils Based on Polarity of Crude Oil Components. Energy and Fuels 2010, 24, 6483–6488.

(5) Mohammadi, A. H.; Eslamimanesh, A.; Gharagheizi, F.; Richon, D. A Novel Method for Evaluation of Asphaltene Precipitation Titration Data. Chem. Eng. Sci. 2012, 78, 181–185.

(6) Rogel, E.; Leon, O.; Contreras, E.; Carbognani, L.; Torres, G.; Espidel, J.; Zambrano, A. Assessment of Asphaltene Stability in Crude Oils Using Conventional Techniques. Energy & Fuels 2003, 17, 1583–1590.

(7) Wang, J. X.; Creek, J. L.; Buckley, J. S. Screening for Potential Asphaltene Problems. In SPE Annual Technical Conference and Exhibition; Society of Petroleum Engineers: San Antonio, 2006.

(8) Cheng Xing, Robert W. Hilts, and John M. Shaw, Sorption of Athabasca Vacuum Residue Constituents on Synthetic Mineral and Process Equipment Surfaces from Mixtures with Pentane, Energy & Fuels, 2010,24(4) 2500-2513.

(9) Amundaraín Hurtado, J. L.; Chodakowski, M.; Long, B.; Shaw, J. M. Characterization of Physically and Chemically Separated Athabasca Asphaltenes Using Small-Angle X-Ray Scattering. Energy & Fuels 2011, 25, 5100–5112.

(10) Pourmohammadbagher, A.; Shaw, J. M. Excess Enthalpy and Excess Volume for Pyridine + Methyldiethanolamine and Pyridine + Ethanolamine Mixtures. J. Chem. Eng. Data 2013, 58, 2202–2209.

SIMULATING THE EFFECT OF CHANGING THE EFFLUENT PATH OF OMAR-BEK DRAIN IN THE DAMIETTA BRANCH WATER QUALITY

Mohamed Mostafa[1] *and Robert W. Peters*[2]
[1]*Postdoctoral Research Associate and* [2]*Professor of Environmental Engineering*
Department of Civil, Construction and Environmental Engineering
University of Alabama at Birmingham
1075 13th Street South
Birmingham, Alabama 35294-4440

Abstract

The Omar-Bek drain represents the main source of water pollution along the Damietta branch of the Nile River, Egypt, where it receives about 600,000 m^3/day of untreated domestic, agricultural, and industrial wastes. The main purpose of this research is to effectively manage water quality at the Damietta branch, involving a comparison between the existing situation and two proposed scenarios. The first scenario involved changing the effluent path of the Omar-Bek drain to another drain called "Main Western drain". The second scenario centered around improving water quality at the Omar-Bek drain by reusing textile wastewater in the manufacture of cement bricks, constructing a WWTP with a design capacity of more than 150,000 m^3/day, and improving water quality at this drain by increasing the efficiency of WWTPs that daily discharge about 60,000 m^3 of partially treated wastewater to the drain. A new river pollutant (RP) modeling has been created using MATLAB software to simulate pollutant transport in the Damietta branch. This software has been created using mass balance principles and contaminant exponential decay equations. River pollutant (RP) modeling is used to calculate water quality parameters such as total organic carbon (TOC), chlorides (Cl^-), biological oxygen demand (BOD), chemical oxygen demand (COD), total dissolved solids (TDS), total suspended solids (TSS), dissolved oxygen (DO), pH, and temperature. The current situation and the two proposed scenarios were simulated using the RP modeling. It was concluded that the Omar-Bek drain has no significant effect on the Damietta branch water quality. As a result, it is recommended not to change the path of the drain, along with preventing the direct discharge of domestic wastewater to the Omar-Bek drain and improving the efficiency of the existing WWTPs discharging to the drain.

Introduction and Background

The longest river in the world, the Nile River measures 6,700 km (4,163.2 mi) long and runs from Lake Victoria in east central Africa to the Mediterranean Sea in northeastern Egypt. The river travels for a 950 km through Egypt, starting from Aswan in the south and ending at Cairo in the north, where it separates into two branches, Rosetta and Damietta branches, which form the Nile delta, as shown in Figure 1. The Damietta branch extends about 242 km (150.37 miles) from downstream of Delta Barrages to the Mediterranean Sea (Ezzat and Reham, 2012) and has an average velocity of about 0.35 m/s and an average depth of about 3.25 m (Drews and Weiqing, 2010; Mostafa, 2014). The top width of the Damietta branch ranges from 200 to 250 m, and the sinuosity ranges from 2.4 to 2.9 (Moussa and Medhat, 2007; Mostafa, 2014). At a distance of 129 km downstream of Delta Barrages, the Damietta branch receives discharge from the Omar-Bek drain. This drain, located in El-Gharbia governorate, starts from Zefta City and ends at Samanoud City and receives pollution loads from domestic, agricultural, and industrial wastes discharged from villages, cites, and drains located along its path (Ezzat and Reham, 2012; Mostafa, 2014).

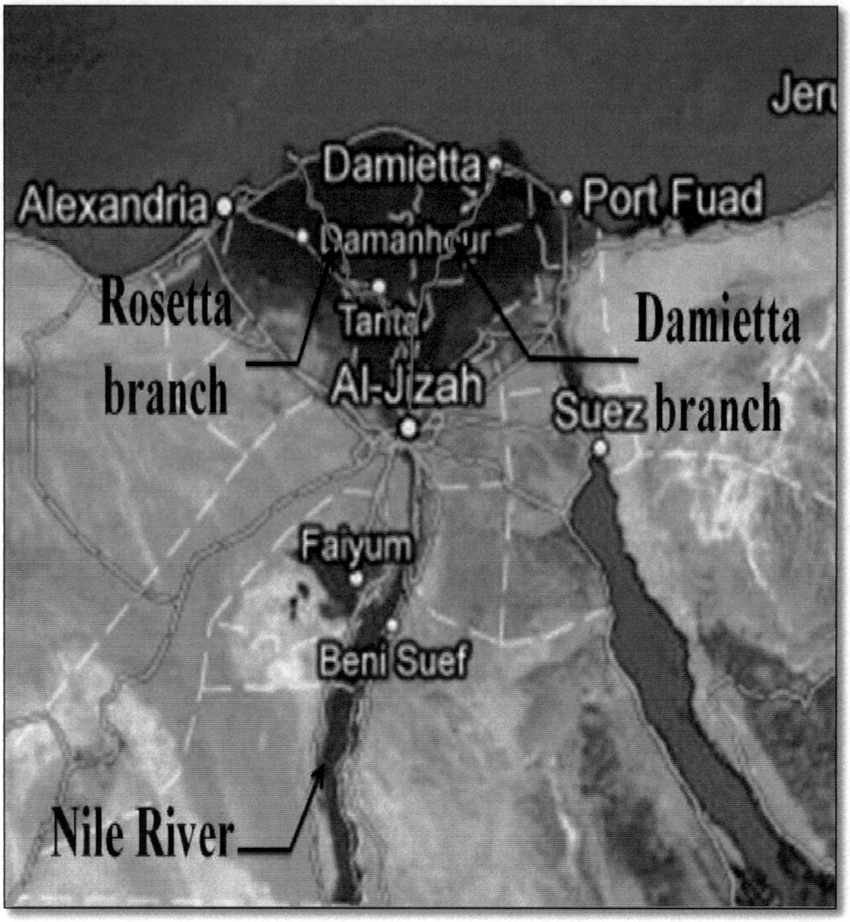

Figure 1. Egyptian Nile valley

Several studies have been conducted to address water quality at the Damietta branch. In a 2010 study, Abdo collected water samples from seven stations along the branch, starting from El-Kanater El-Khyria Barrage and ending at Damietta estuary, and concluded that the change in water quality at the Damietta branch resulted from thermal pollution, climatic conditions, agricultural wastes stored behind Faraskour Dam, and sewage wastes at El-Serw City. In a 2012 study, Ezzat and Reham investigated the effect of the Omar-Bek drain on the Damietta branch water quality. Water samples were collected at the discharge point of the Omar-Bek drain and downstream and upstream of the discharge point. The water quality index was calculated, and the results led to classifying the surface water quality upstream of the Omar-Bek drain as good and the surface water downstream of the drain as marginal. It was also determined that the concentrations of heavy metals (aluminum, iron, selenium, arsenic, zinc, copper, manganese, barium, chromium, cadmium, tin, lead, nickel, cobalt, and vanadium) at the Omar-Bek drain and upstream and downstream of the drain did not exceed the permissible limits specified in Egyptian law 48/1982 (NWRC, 1995).

Pollution prevention (P2) assessment was conducted in one of the textile industries located in Giza governorate, Egypt. The on-site P2 assessment started with an initial meeting between the assessment team and the company representative. Company representatives reviewed the needs of the company and reviewed all available information on the manufacturing cycle within the facility. After the meeting, a company representative led the assessment team for a walk-through tour of the plant. During the walk-through, the assessment team verified the processes contributing to a waste stream, and collected mass balance information. The assessment team found that approximately 660,000 gallons/year of wastewater are generated from dyeing and finishing stages. The assessment team also observed that the wastewater is discharged without treatment to an agricultural drain which discharges directly to the Nile River. Wastewater samples were collected to be used in the manufacture of the cement bricks. Different samples were prepared and tested to evaluate the effect of using wastewater on the properties of cement bricks. Finally, the assessment team prepared and submitted a report to the company. This report discussed applicable technology that can be used to reuse wastewater.

This research was conducted to manage water quality at the Damietta branch; in this study, a comparison was made between the current situation and the situation involving changing the effluent path of the Omar-Bek drain to the Main Western drain. Both situations were simulated with the use of river pollutant (RP) modeling. The study area encompassed about 119 km (73.94 miles) of the Damietta branch and extended from upstream of the Omar-Bek drain to Faraskour City. Figure 2 depicts a map of the study areas.

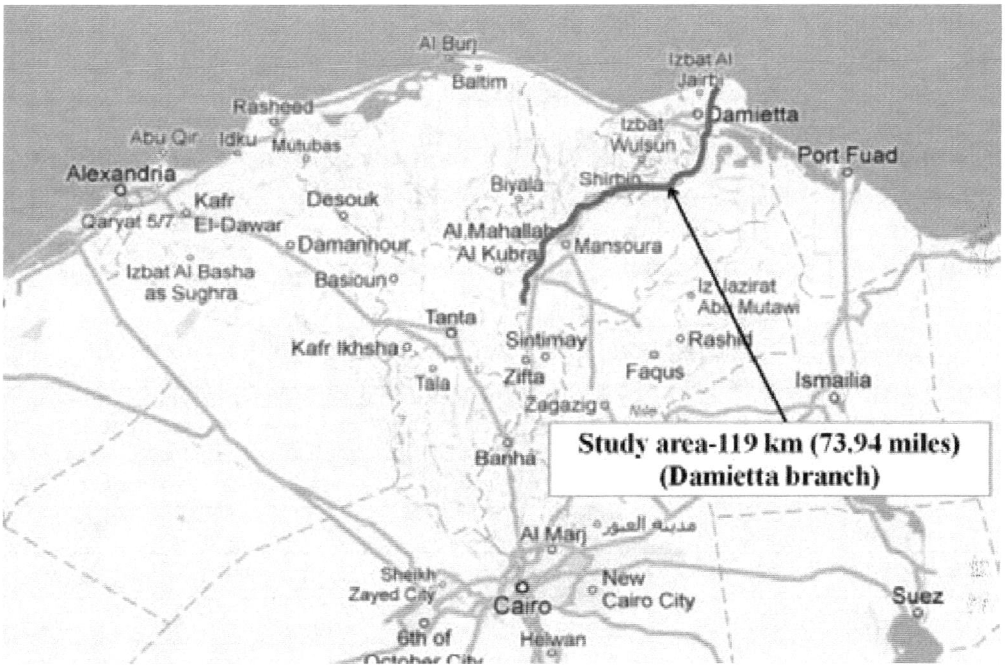

Figure 2. Study area at the Damietta branch

Results and Discussion

A paired-samples *t*-test was performed to evaluate the difference between the means of the samples made with clean water and textile wastewater. The null hypothesis (H_0) states that there are no differences between the means of the samples, while the alternate hypothesis (H_1) states that there are differences between the means of the samples. As shown in Table 1, the *t*-statistic values were greater than the critical *t*-value for all the tests. The parameter *p*-values were also found to be significantly less than 0.05, as shown in Table 1. Consequently, there was a significant difference between the means of the samples; in this case, the null hypothesis was rejected in favor of the alternative hypothesis. Using the textile wastewater in the manufacture of cement bricks will improve the mechanical properties of the bricks.

Table 1. Test for bricks and *t*-test (paired two-samples for comparison of means).

Test	Clean water	Textile wastewater	EPA limit	*t*-statistic	*t*-critical two-tail	*p*-value	Hypothesis
Compressive strength	170.65	174.57	> 140 kg/cm^2	27.08	±4.302	0.00136	H_1
Water absorption	227	207.7	< 240 kg/m^3	-58	±4.302	0.000297	H_1
Flash point	97°C (205°F)	94°C (199°F)	60°C (140°F)	39.70	±4.303	0.0006342	H_1
Corrosivity, pH	12.04	11.34	pH < 2.0 or > 12.50	10.05	±4.302	0.0097511	H_1
Sulfide, Total	85 mg/kg	110 mg/kg	500 mg/kg	-26	±4.303	0.001476	H_1
Cyanides, Total	ND	< 1.0 mg/kg	250 mg/kg	N/A	N/A	N/A	N/A

For the first scenario (changing the effluent path of the Omar-Bek drain), the data from the Omar-Bek drain were excluded during the running of the modeling. Water flow and characterization at the Omar-Bek drain for the current situation and the second proposed scenario, are presented in Table 2. The RP modeling was then used to simulate pollutant transport for the three scenarios.

Table 2. Water flow and characterization at the Omar-Bek drain.

Term	Current situation	Second scenario
Flow, m^3/day	600,000	600,000
BOD, mg/L	55.0	23.20
COD, mg/L	97.0	44.65
TOC, mg/L	1.80	1.62
TDS, mg/L	700	608.65
TSS, mg/L	160	30.30
DO, mg/L	3.90	4.325
Cl^-, mg/L	130	122.40
pH, dimensionless	7.50	7.90

Second scenario: improving water quality at the Omar-Bek drain

The simulations of the current situation and the two scenarios are shown in Figures 3 through 7. Figure 3 shows the predicted DO concentration for the three scenarios along the study area. Applying the proposed solutions will have a negligible effect on the concentration of the DO, where applying the first scenario is expected to increase the DO concentration downstream the Omar-Bek drain from 6.0 to 6.10 mg/L, while applying the second is expected to increase the DO concentration from 6.0 to 6.05 mg/L (see Figure 3). Consequently, the Omar-Bek drain slightly reduced the DO concentration in the Damietta branch. Figure 4 shows the predicted BOD concentration for the current situation and the two proposed scenarios along the study area. Applying the first scenario will reduce the BOD concentration from 3.33 to 2.0 mg/L downstream of the Omar-Bek drain; while applying the second scenario will reduce the BOD concentration from 3.33 to 2.50 mg/L (see Figure 4). Consequently, applying both scenarios has a minor impact on reducing the BOD concentration in the Damietta branch.

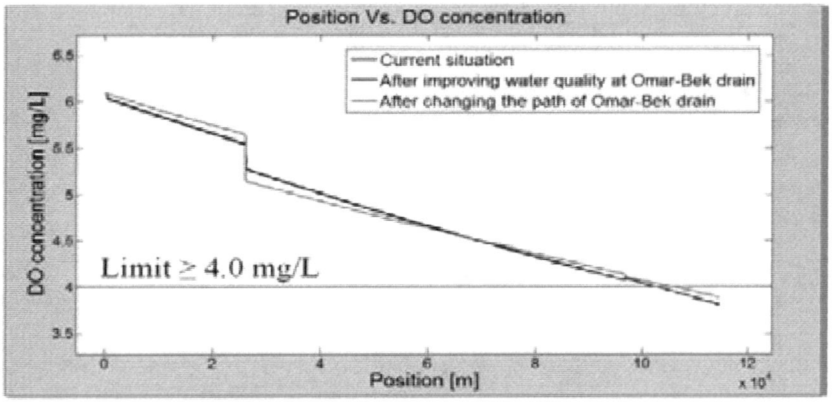

Figure 3. DO Concentration along the Damietta Branch for Current Situation and Proposed Solutions

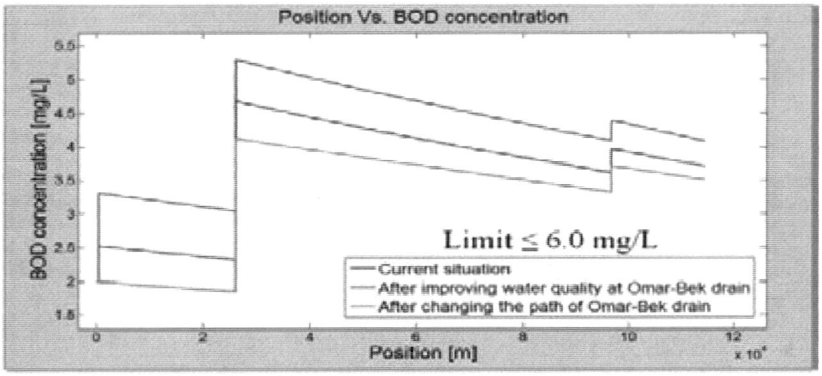

Figure 4. BOD Concentration along the Damietta Branch for Current Situation and Proposed Solutions

The water quality standard for TDS is 500 mg/L (Egyptian Law 48/1982 and EPA Standards) (NWRC, 1995; EPA, 2007). Applying the first scenario is expected to decrease the TDS concentration downstream the Omar-Bek drain from 202.60 to 190.0 mg/L, while applying the second is expected to decrease the DO concentration from 202.60 to 200.0 mg/L (see Figure 5). Consequently, the discharge from the Omar-Bek drain did not adversely affect TDS concentration in the Damietta branch and applying the proposed solutions will have a slight effect on the concentration of the TDS in the Damietta branch. Figure 6 shows the predicted Cl^- concentration for the three scenarios along the study area. Applying the second scenario is ineffective because the Omar-Bek drain has only a slight effect on increasing Cl^- concentration in the Damietta branch. The second scenario also has a minor impact on reducing the Cl^- concentration downstream of the Omar-Bek drain, where the Cl^- concentration downstream the Omar-Bek drain is expected to decrease from 34.41 mg/L to34.30 mg/L (see Figure 6). After the branch received discharge from the Omar-Bek drain, the TOC concentration at the Damietta branch increased to 0.24 and 0.23 mg/L for the current situation and for the second scenario, respectively (see Figure 7). In the case of the first scenario, the TOC concentration upstream of the Omar-Bek drain is approximately equal to the TOC concentration downstream of the drain.

Consequently, applying the proposed solutions proved ineffective because the Omar-Bek drain only slightly increases TOC concentration in the Damietta branch.

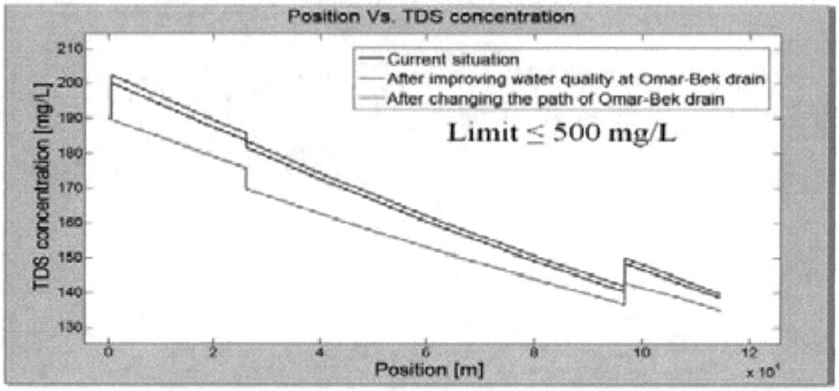

Figure 5. TDS Concentration along the Damietta Branch for Current Situation and Proposed Solutions

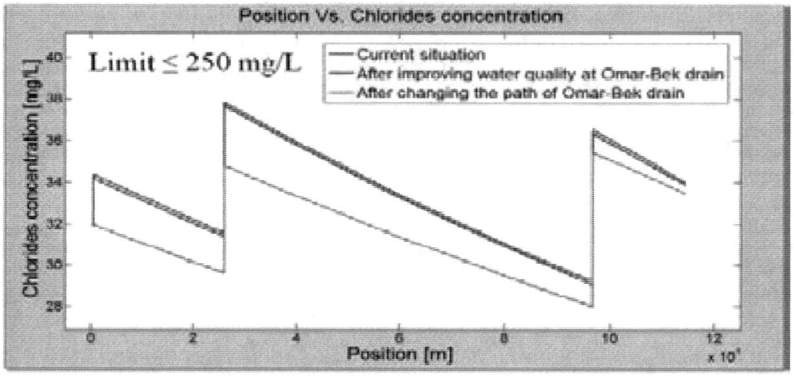

Figure 6. Cl^- Concentration along the Damietta Branch for Current Situation and Proposed Solutions

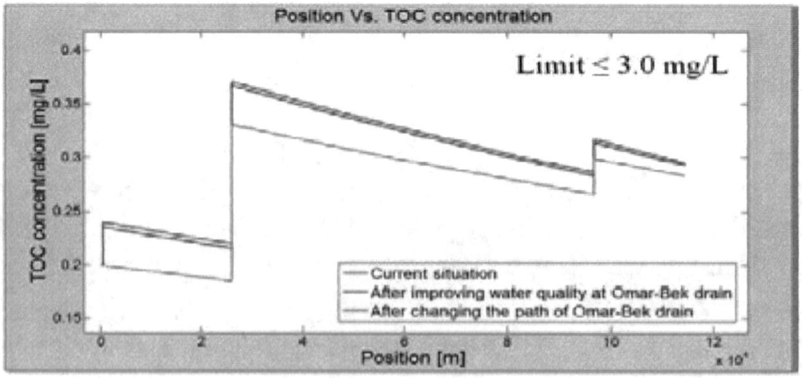

Figure 7. TOC Concentration along the Damietta Branch for Current Situation and Proposed Solutions

Summary and Conclusions

The RP modeling showed that discharge from the Omar-Bek drain does not significantly affect Damietta branch water quality. The modeling result also showed that improving water quality at the Omar-Bek drain by reusing textile wastewater in the manufacture of cement bricks, constructing a

WWTP, and improving the efficiency of existing WWTPs is more preferable than changing the path of the drain. Another recommendation for improving water quality at the Omar-Bek drain involves installing sewage networks to prevent the direct discharge to the drain. The total cost of changing the effluent path of the Omar-Bek drain could reaches $25 million. Because the Omar-Bek drain does not significantly affect Damietta branch water quality and because the cost of changing the path is high, it is recommended that the path of the drain not be changed and that these funds be directed toward more complex problems such as improving the efficiency of the existing WWTPs discharging to the drain.

List of References

Abdo, M.H., (2010). "Environmental and Water Quality Evaluation of Damietta branch, River Nile, Egypt", National Institute of Oceanography and Fisheries, Inland Waters and Aquaculture Branch, ***African J. Biol. Sci., 6*** (2): 143-158.

Drews C., H. Weiqing, (2010). "Dynamics of Wind Setdown at Suez and the Eastern Nile Delta", PLOS ONE Community blog, 5(8), Boulder, CO, USA.

Environmental Protection Agency (EPA), (2007). "Drinking Water Standards and Health Advisories Table". San Francisco, CA.

Ezzat, S.M, and M.E. Reham, (2012). "Omar Bek Drain Water Quality and its Impact on Damietta Branch, River Nile-Egypt", ***American-Eurasian J. Agric. & Environ. Sci., 12*** (4): 472-483.

Mostafa, M. (2014) Modeling of Pollutant Transport in the Nile Delta Egypt, PhD. Dissertation, Department of Civil, construction, and Environmental Engineering, University of Alabama at Birmingham, Birmingham, AL, USA.

Moussa, A.M.A., and S.A. Medhat, (2007). "Nile River Sediment Transport Simulation (Case Study: Damietta Branch)", ***Eleventh International Water Technology Conference***, IWTC11, Sharm El-Sheikh, Egypt. (March 15-18, 2007)

National Water Research Center (NWRC), (1995). "River Nile Protection and Development Project Phase II", Environmental Pollution and Legislative Regulations (Law 48. 1982 & Decree 8, 1993), Ministry of Public Works and Water Resources, Report No. A3010901 – 9508 – 01 – FR.

414871 Biocompatible Ionic Liquid As Inhibitor for Qatari Natural Gas Hydrates

Abstract:

The formation of stable gas hydrates in the production and transmission pipelines in oil and gas industries can cause operational and safety hazards, as well as production and massive economical loss. Due to the vast development in deep water activities carried out in many locations all over the world including Qatar, flow assurance has become one of the major and critically important challenges in overcoming hydrates problem [1]. Thereby, search for new class of chemicals know as gas hydrate inhibitors which are economic and used in low dosages are the centre of attraction to avoid hydrate formation [2].

In this work, hydrate equilibrium curves have been obtained for a Qatari natural gas type mixture (QNG) using a bench top reactor and gas hydrate autoclave. The hydrate inhibition effect of a biocompatible ionic liquid, choline chloride (ChCl) has also been tested at lower concentrations (1 and 5 wt%) and found little compared to conventional thermodynamic inhibitors. Moreover purging of nitrogen into the system to make the QNG mixture N_2 rich up to approximately (0.5 QNG+0.5 N_2) has also been studied in presence of 1 and 5 wt% of ChCl. It has been found that the presence of N_2 and ChCl together enhance the performance of thermodynamic inhibition. The obtained results suggest that the synergistic inhibition effect of (N_2+ChCl) could be an alternative economic strategy for the inhibition of natural gas hydrates.

Further, the obtained experimental for QNG were compared to each other and to the predicted curves using commercial software (HydraFLASH) and Katz gas gravity method. The possible reasons of the discrepancies between the experimental and predicted values were discussed in terms of the mixture composition uncertainty and the limitation of software prediction power for complex mixtures like QNG.

[1] M. Tariq, D. Rooney, E. Othman, S. Aparicio, M. Atilhan, M. Khraisheh, Gas hydrate inhibition: a review of the role of ionic liquids, *Ind. Eng. Chem. Res.* 53 (2014) 17855 – 17868.

[2] M. A. Kelland, History of the development of low dosage hydrate inhibitors, *Energy Fuels* 20 (2006) 825 – 847.

419872 Prediction of Pipeline Erosion Uncertainty for Scale-up

Abstract:

Prediction of pipeline erosion uncertainty for scale-up

Wei Dai, Ravi Nukala, Selen Cremaschi

[a]Russell School of Chemical Engineering, The University of Tulsa, 800 S Tucker Drive, Tulsa,OK, USA

selen-cremaschi@utulsa.edu

Abstract

In this study, a dimensional analysis is used for quantification of erosion-rate prediction uncertainty within pipelines that transport particles in multiphase gas-liquid flows. The transport of solids in multiphase flows is common practice in energy industries because of the unavoidable extraction of solids from oil and gas bearing reservoirs either onshore or offshore sites. The safe and efficient operation of these pipelines requires reliable estimates of erosion rates, and production rates are generally limited to keep the effects of erosion at acceptable levels. Erosion in pipelines is defined as the material removal from the solid surface due to solid particle impingement. The phenomena that leads to this type of erosion, especially in multiphase flow systems, is very complex and depends on many factors including fluid and solid characteristics, the pipeline material properties and the geometry of the flow lines.

Given this complexity, most of the modeling work in this area focuses on developing empirical or semi-mechanistic models. For example, Oka et al. (2005) developed their erosion model using particle impingement in air with empirical constants based on particle properties and hardness of the target materials. Their model is one of the most commonly cited in the literature. Another semi-mechanistic model called 1-D SPPS (Zhang, 2007), which is widely used for predicting erosion rates by the oil and gas industry, was developed with several empirically estimated parameters, like the sharpness factor of particles, Brinell hardness and the empirical constants in the impact angle function. These empirical parameters are calculated using experimental observations. However, the experimental data used in these calculations and also for model validation and uncertainty quantification are, for the most part, collected in small pipe diameters (from 2 to 4 inches). These small pipe sizes do not coincide with the field conditions, where the pipe diameters generally exceed 8 inches. Hence, the predictions of erosion models are routinely extrapolated to conditions where experimental data or even operating experience is not available, and the estimation of erosion-rate prediction uncertainty becomes crucial especially for systems too-costly to fail. The goal of this study is to develop a systematic approach to estimate erosion-rate prediction uncertainty for extrapolations. To achieve this goal, the erosion-rate model discrepancy, which is defined as the difference between experimental erosion rates and the corresponding erosion rate predictions, is modeled using Gaussian Process Modeling (GPM, Rasmussen, 2006). We used functions of dimensionless numbers that are relevant to erosion phenomena as the inputs to the GPM. Use of dimensionless numbers as inputs enables the scale-up of uncertainty estimates.

The GPM models erosion-rate model discrepancy as a Gaussian random process, which is defined by mean and covariance functions assuming a multivariate normal distribution (Zhen, 2013). The most likely values of mean and covariance function parameters are determined by Maximum Likelihood Estimation (MLE) using experimental data. The GPM model not only provides prediction in locations where experimental data is not available but also constructs the prediction confidence using the covariance functions through conditional probability distribution.

The approach developed to determine dimensionless groups and the functions as inputs to GPM has four main steps: (1) Identify all possible dimensionless groups that are relevant to erosion phenomena using Buckingham ¦Ð theorem. Here, we considered pipe diameter, particle size, density of particle and flow, viscosity of flow, flow rates, gravitation constant and surface tension as the dimensional variables. Time, length and mass are the three basic units. Therefore, we obtain 67 sets of 8 dimensionless numbers. Because some dimensionless numbers are repeated in different sets, the Buckingham ¦Ð theorem yields 200 distinct dimensionless numbers. Including pipe geometry, particle hardness and sharpness, which are already dimensionless, results in 203 distinct dimensionless numbers. (2) Calculate the correlation between each dimensionless number and model discrepancy, and find the sum of the correlation coefficient values that are greater than 0.5 (these refer to strong correlations) for each set. (3) Select the dimensionless group sets with the five highest correlation sums, and flag their corresponding dimensionless numbers as candidate inputs to GPM. (4) Formulate and solve the optimization problem to select the best function of candidate dimensionless numbers. Here, the best is defined as the function that minimizes the selected performance metric, i.e., area metric.

The performance of different input sets and functions as inputs to GPM is assessed using a modified area metric (Ferson, 2008). Area metric is defined as the integral of disagreement area between the estimated erosion rate and experimental data. A smaller area metric represents a better prediction of the model discrepancy. It can also be used to locate under-prediction regions of the erosion model.

For estimating erosion-rate-prediction uncertainty, we compiled an experimental database of erosion rate measurements from literature. It contains 544 data points in single or multiphase carrier flows. Eighty percent of the data in the database are collected for gas dominated flows (i.e., gas only, annular, mist and churn flow). The experimental database covers a wide range of input conditions resulting in significantly different erosion rates.

We selected the 1-D SPPS model as our case study, quantified its erosion-rate discrepancy using our developed approach. The 1-D SPPS model calculates the maximum erosion by defining how a hypothetical representative particle will impinge the target material. The abrasion caused by this particle is defined by length loss in the target material, and is calculated using the momentum of impingement. The maximum erosion rate model in 1-D SPPS calculates the target material length loss per unit time, and uses a power law correlation of the characteristic impact velocity. The 1-D SPPS model accounts for pipe geometry, size and material, fluid properties (density and viscosity) and rate, and particle sharpness, density and rate. The 1-D SPPS model discrepancy is calculated for 544 experimental data points. The model discrepancies and the corresponding values of the candidate dimensionless numbers are used as inputs to our uncertainty estimation framework. Then, the overall analysis is performed for each flow regime and the corresponding average area metric is calculated. For mist flow, gas velocity, density and pipe diameter are the repeating variables that yielded the dimensionless groups with the smallest averaged area metric value, which was equal to 0.047. For churn flow, particle size, gas density and surface tension are the repeating variables that yielded the dimensionless groups with the smallest averaged area metric value, which was equal to $9.2x10^{-4}$. For slug flow, pipe diameter, gas viscosity and surface tension are the repeating variables that yielded the dimensionless groups with the smallest averaged area metric value, which was equal to $8.1x10^{-4}$. While for annular flow, gas velocity, gas density and surface tension are the repeating variables that yielded the dimensionless groups with smallest averaged area metric value, which was equal to 0.0074. Those dimensionless groups identified in each flow regime provided most influential variables in the quantification of erosion-rate model discrepancy and also suggested possible modeling and experimental improvements involving those variables. Besides, extrapolation of erosion discrepancy prediction is possible based on the identified dimensionless groups in each flow regime.

Our analysis indicate that the use functions of dimensionless numbers as inputs to GPM improves the erosion rate prediction discrepancy and reduces the confidence intervals of uncertainties compared to using dimensional inputs to GPM as evidenced by smaller area metrics. More specifically, compared to GPM results with dimensional inputs, our approach yields a 60% decrease of area metric value in mist flow, 23% decrease in churn, 78% decrease in slug and 12% increase in annular flow with respective best dimensionless groups.

Acknowledgement

This work is supported by the Chevron Energy Technology Company. Discussions and comments from the Haijing Gao, Gene Kouba and Janakiram Hariprasad of Chevron and Brenton McLaury, Siamack Shirazi of E/CRC at the University of Tulsa were highly acknowledged. <>Reference

Ferson, S., Oberkampf, W. L., and Ginzburg, L., 2008, Model Validation and Predictive Capability for the Thermal Challenge Problem, Computer Methods in Applied Mechanics and Engineering, Vol. 197, No. 29-32, pp 2408-2430.

Jiang, Z., Chen, W., Fu, Y., and Yang, R., 2013, Reliability-Based Design Optimization with Model Bias and Data Uncertainty, SAE International.

Oka, Y. I., Okamura, K., and Yoshida, T., 2005, Practical estimation of erosion damage caused by solid particle impact: Part 1: Effects of impact parameters on a predictive equation, Wear, 259(1-6), page 95-101.

Rasmussen, C.E. and Williams, C.K. I., 2006, Gaussian Processes for Machine Learning, The MIT Press.

Zhang, Y., Reuterfors, E.P., McLaury, B.S., Shirazi, S.A., and Rybicki, E.F., 2007, Comparison of Computed and Measured Particle Velocities and Erosion in Water and Air Flows, Wear, 263.

421579 Study on Poly(styrene-co-alkyl maleimide) As a Flow Improver for the Model Crude Oil with Heavy Asphaltenes

Abstract:

*Corresponding Author's E-mail: kcao@che.zju.edu.cn

This study was supported by the National Natural Science Foundation of China through Project No. 21176217.

The waxes and asphaltenes in crude oils are detrimental to pipeline transportation and processing due to the significant solubility decrease of waxes and asphaltenes with the depression in temperature or variation of other conditions, causing formation and deposition of stable wax crystals and asphaltene particles. Polymeric flow improvers are often widely used to solve this problem by retarding the wax crystallization and inhibiting the asphaltene aggregation. In our previous work, model waxy crude oils without and with low content of asphaltenes (0.1 wt %) have already been investigated, which indicates that the various molecular structure of the flow improvers have significant influence on the effectiveness of flow improving.

In present work, a synthesized flow improver, poly (styrene-co-octadecyl maleimide) is studied intensively for the performance in improving flowability of the model waxy crude oil with much higher content of asphaltenes (3 wt %). The characteristics of model crude oil have been essentially altered by the great amount of asphaltenes present and are quite different with that of the aforementioned two kinds of model crude oils. As the asphaltenes have been found to be incompatible with the waxes, they can readily precipitat followed by further aggregation forming large asphaltene particles and finally sediment especially at such high concentration. Hence the model crude oil with high content of asphaltenes is a heterogeneous system other than a homogeneous system as the model crude oil without or with low content of asphaltenes is. Therefore, the flow improving performance and mechanism of flow improvers may be different.

Flow improvers with a variety of structures are derived from the precursor, poly (styrene-co-maleic anhydride). The influence of various molecular structure, including molecular weights, maleic anhydride content of precursor, imidization degree and block structure on the effectiveness of flow improvers is thoroughly explored herein. The influence is interpreted in aspects of the yield stress and viscosity of the model crude oil, the stability of asphaltene particles in matrix, morphology of asphaltenes and wax crystallization temperature by the methods of rheometer, particle stability analysis through Turbiscan Lab Expert, microscopy and differential scanning calorimetry, respectively. Compared to untreated model crude oil, the viscosity and yield stress of crude oils with flow improvers are decreased considerably. The asphaltene particles are stabilized, better dispersed and the sedimentation is retarded. The wax crystallization temperature is depressed as well. Hence, poly (styrene-co-octadecyl maleimide) with an optimal structure can potentially be a kind of efficient flow improver for heavy crude oil with high content of asphaltenes and waxes.

(425a) Effectiveness of Pyrrolidinium Based Ionic Liquids in Flow Assurance: Preventing Hydrates Formation in Offshore Sub-Sea Pipelines

Qureshi, M. F., Qatar University
Atilhan, M., Qatar University
Tariq, M., Qatar University
Khraisheh, M. Sr., Qatar University

The formation of hydrates within the offshore subsea pipelines is a major challenge to flow assurance in oil and gas industry. Annually, the industry spends large amount of money on thermodynamic inhibitors like methanol and methyl ethylene glycol (MEG) to prevent hydrates formation. However, large quantity (>30 wt %) of MEG and methanol is required for this process, causing disposal and corrosion issues. MEG is also very expensive and methanol is dangerous to use in offshore facilities due to its high flammability. This has shifted the interest of researchers towards the use of ionic liquids (ILs) to prevent hydrates formation in sub-sea pipelines. Thus, the aim of this work is to test the effectiveness of pyrrolidinium based ILs in preventing hydrates formation and check for exhibition of any dual inhibition behaviour by these ILs.

We have tested three pyrrolidinium based ILs namely: 1-methyl-1-propyl pyrrolidinium chloride, 1-methyl-1-propyl pyrrolidinium triflate and 1-methyl-1-propyl pyrrolidinium dicyanamide for their effectiveness in hydrate inhibition of a synthetic quaternary gas mixture containing: methane (85%), ethane (10%), nitrogen (2.5%) and carbon dioxide (2.5 %). It is evident that the cation of tested ILs is fixed as ''1-methyl-1-propylpyrrolidinium'' and anions are changed which will help us to understand the role of anions, if any, in inhibition performance of the tested ILs. These tests were conducted using 1 and 5 wt% concentrations of each IL sample on a rocking cell assembly by PSL Systemtechnik GmBH at different pressures ranging from (40-120) bars.

According to the tests results, the pyrrolidinium based ionic liquids were able to prevent hydrates formation by providing a temperature shift of 1.8 at low pressures and by delaying hydrates formation time by almost one hour at high pressures. These results also confirm the dual inhibition behaviour of pyrrolidinium based ionic liquids, as they were able to show both thermodynamic and kinetic inhibition effects simultaneously. However, no significant effect of anion is observed. The results were also compared with standard industrial thermodynamic inhibitor methanol and the trends are discussed.

413703 Investigation of the Potential of the ICP-Qqq for the Analysis of Petroleum Fractions

Abstract:

Intensive studies have been dedicated to trace and ultra trace metals determination in petroleum products in the recent years. The need of the petroleum industry in this field is related to exploitation activities and exploration : contamination during oil production and refining (e.g. prevention of catalyst poisoning, corrosion and pollution control), and geochemical characterization of source rocks and basins in order to identify geochemical biomarkers in oil-oil or oil-source rock correlation. The existence of metals in fossil fuels was first established by Alfred Treibs in the 1930s [*1*]. Trace metals are incorporated into petroleum fractions as organometallic compounds (e.g. geoporphyrins) initially present in the crude oil [2] or added during the various processes of refining [3], but also as inorganic components originating from formation and production of water (e.g. contamination of Ba and As), and mineral matter (e.g. trace elements in clay minerals) [*4*]. Their concentration are always in the mg/kg or even µg/kg levels but low levels of certain metals can have drastic effects as a contaminant. Conventionally, ICP techniques are used [5] and allows sufficient detection limit for most of the cases of the petroleum industry. The inherent high sensitivity of ICP-MS detection together with isotopic ratio capabilities opens new fields of applications in petrochemistry or geochemistry of petroleum products, but the introduction of organic substances is not that easy as this technique was not initially designed for organic samples analysis and specific configuration of sample introduction systems are required in order to minimize organic solvent load into the ICP plasma. Moreover, specific polyatomic interferences are disturbing the spectra due to the massive presence of carbon and oxygen within the plasma. These interferences are solved with the use of a collision/reaction chamber or the use of high resolution ICP/MS [6].

The recent commercialisation of a triple quad ICP/MS by Agilent, specifically dedicated for heavy matrices analysis might be an interesting alternative to solve polyatomic interferences with petroleum fractions and attain lower detection limit. This contribution will present the result obtained with such a system compared to high resolution ICP/MS for various element and matrices tested.

[1] Treibs, A., "Porphyrins in coal", Anal. Chem., Vol. 520, 1935, pp. 144-151.

[2] Caumette G., Lienemann C.-P., Merdrignac I., Bouyssiere B., Lobinski R., Element speciation analysis of petroleum and related materials, J. Anal At. Spectrom., 24, (2009), 263-276

[3] Chainet F., Le Meur L., Lienemann C.-P., Ponthus J.,Courtiade M., Donard O.F.X, Characterization of silicon species issued from PDMS degradation under thermal cracking of hydrocarbons: Part 2 -Liquid samples analysis by a multi-technical approach based on gas chromatography and mass spectrometry, Fuel, 116, (2014), 478-89

[4] Filby, R.H., "Origine and nature of trace elements species in crude oil, bitumens and kerogen: implications for correlation and other geochemical studies", in: J. Parnell (Ed.), Geofluids: Origin, Migration and Evolution of Fluids in Sedimentary Basins, Geol. Soc. Sp. Pub., Vol. 78, 1994, pp. 203-221.

[5] Sánchez R., Todolí J.-L; Lienemann C.-P., Mermet J.-M., Determination of trace elements in petroleum products by Inductively Coupled Plasma techniques: a critical review, Spectro Chimica Acta Part B, 88, (2013), 104-26

[6] Pohl P., Dural D., Vorapalawut N., Merdrignac I, Lienemann C.-P., Carrier H., Grassl B., Bouyssiere B., Lobinski R., Multielement molecular size fractionation in crude oil and oil residue by size exclusion microchromatography with high resolution inductively coupled plasma mass spectrometric detection (HR ICP MS), J. Anal At. Spectrom., 25, (2010), 1974-7

431761 Towards a Generalized Molecular Model for Upstream and Downstream Applications

Abstract:

Phase equilibrium experiments, Fourier-transformed infrared (FT-IR) spectroscopy measurements, molecular dynamic (MD) simulations and the Perturbed Chain form of the Statistical Associating Fluid Theory (PC-SAFT) were used to understand the interactions in water + alkane, alcohol + alkane, alcohol + aromatic and water + alcohol binary systems. Modeling water + alkane mixtures has been a long standing challenge. Using molecular simulation, it was found that an effective water Lennard-Jones energy of 220 K is required to match the experimental water solubility in TraPPE alkanes. A similar value of dispersion energy was required when PC-SAFT was applied. These numbers are significantly higher than that used in most of the water simulation models (78.2 K for SPC/E). The reason behind such discrepancy is yet to be found. Nevertheless, the application of the new SAFT water model was extended to predict water content of sweet and sour natural gas mixtures in equilibrium with an aqueous or a hydrate phase. In general, alcohols and water exhibit similarities in terms of structure and physical interactions. As a result, studying the interactions in alcohol + alkane systems might provide insight into the water + alkane problem. The Polar PC-SAFT equation of state was used to model the system vapor-liquid equilibrium (VLE), activity coefficients and extents of association. The model showed a good agreement with spectroscopic and simulation data on the fraction of free alcohol monomers. Moreover, the alcohol model was used to predict the thermodynamics of weak hydrogen bonds formed between alcohols and the π-electrons of the aromatic rings. Results were compared with spectroscopic data and showed an acceptable agreement. Finally, using the knowledge gained in understanding water interactions, cluster formation in alcohol + water binary systems was studied using molecular theory and MD simulations.

421640 A Series of Poly(styrene-co-alkyl maleimide) As Versatile Flow Improvers for the Crude Oils with Asphaltenes and Waxes

Abstract:

*Corresponding Author's E-mail: kcao@che.zju.edu.cn

This study was supported by the National Natural Science Foundation of China through Project No. 21176217.

Crude oils are complex mixtures of hydrocarbons that contain polar components such as asphaltenes and nonpolar long-chain waxes. Within the decades the massive consumption of crude oils results in the deterioration of quality for crude oil resources worldwide, indicating the increasing contents of asphaltenes and waxes in crude oils. Once the conditions such as temperature or pressure are changed in the reservoir or pipeline, asphaltenes and waxes precipitate out of crude oil matrix with the dramatic decreasing solubility. The aggregation of long-chain wax solids and asphaltenes particles can lead to deposition at the pipe walls, reducing crude oil flowing ability and potentially clogging the pipeline. Such a problem is particularly severe in cold regions and offshore deep water.

Flow improvers, which are usually polymeric additives, have been widely utilized to avert the clogging problem for the crude oils. Most of previous researches have separately focused on inhibiting asphaltene aggregation and modifying wax crystallization. Few has study the simultaneous effects on asphaltene and wax. In this work, the novel flow improvers, poly(styrene-co-octadecyl maleimide), with various molecular structures were prepared. The precursor, poly(styrene-co-maleic anhydride), was first synthesized through RAFT polymerization, by which the various structure, such as different molecular weights, alternate sequence, random sequence with different maleic anhydride contents and block structure, can be readily attained. Then the precursors were amidated with n-octadecyl amine and subsequently imidized to obtain the desired flow improvers with different imidization degrees by controlling reaction time. The influence of molecular structures of flow improvers on the flow improving effectiveness for various kinds of model crude oils have been intensively examined by the methods of rheology, cross-polarized microscopy, differential scanning calorimetry and stability analysis.

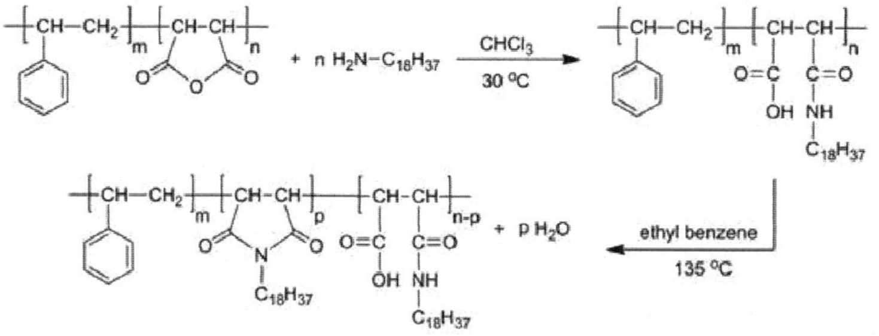

Scheme 1. The imidization processes for obtaining the designed flow improvers

For model waxy crude oils without asphaltenes, the results indicated that adding a small amount of the flow improvers can significantly reduce the crystallization temperature, the number and size of waxy crystals, and the yield stress of the model oil. Moreover, it has been found that the effectiveness of flow improvers is significantly a ected by the imidization degree and maleic anhydride content. With increasing imidization degree, more maleamic acid groups can be converted into maleimide groups, which promotes the compatibility between nonpolar wax and flow improvers.

For model waxy crude oils with asphaltenes, with the addition of the flow improvers, the wax crystals become fewer, smaller, and more dispersed in the oils. Asphaltene particles are better dispersed, and the crystallization temperature and the yield stress can be reduced considerably. It is consistent with results found for waxy crude oils without asphaltenes that the crystallization temperature can be reduced with increasing imidization degree. Consistent with the preceding results, it could be seen that the addition of flow improvers decreases the number and size of wax crystals more significantly with the imidization degree increasing. However, contrary to that without asphaltenes, there is an optimal medium imidization degree for dispersing the asphaltene particles and best depressing the viscosity and yield stress of waxy crude oils with asphaltenes. With the addition of asphaltene, the model waxy oils show higher polarity than before. The flow improvers with medium imidization degree possess polar moieties such as COOH and NH, which are likely to form hydrogen bonds with the asphaltenes containing OH and NH groups, which can benefit the dispersion of asphaltenes. Furthermore, the investigation of those flow improvers for model waxy crude oil with higher content of asphaltenes is also carried out, in which the stability of asphaltene particles have been explored using Turbiscan Lab Expert.

429124 High Pressure Calorimetry and Thermal Analysis Applied to Flow Assurance Related Topics

Abstract:

With the increasing number of deep offshore operations, the operators and the service companies have to solve more and more complex technical challenges related to flow assurance. In both gas transportation and oil exploration, gas hydrate formation can modify the flow characteristics and in some cases lead to blockages and catastrophic events. Thus it is important to prevent the formation of hydrates and ensure optimal operational conditions.

The usual way to determine the thermodynamic conditions of the formation of hydrates in drilling mud formulations is to use a PVT cell with visual observation and simultaneous temperature and pressure measurements. This technique requires heavy instrumentation and often cannot be used if solid particles are present in the formulation.

Another crucial topic for the flow assurance is related to the precipitation of paraffins contained in crude oils, which can occur in reservoirs, pipelines and process equipment resulting in a reduced production and/or a blocked line. To evaluate the possible wax precipitation of a given fluid, the wax appearance temperature (WAT) has to be determined. Most of the studies about WAT are carried out under atmospheric pressure. However, WAT determination under experimental conditions close to real operating ones (particularly under pressure) is a key point to predict the crude oils behavior.

High pressure calorimetry is another way to determine compositions, dissociation enthalpies and heat capacities of hydrates and to provide a detailed characterization of the crude oil in order to prevent the wax precipitation. Different examples of how this technique can be used to determine the thermodynamic properties and the kinetics of hydrate formation, as well as the WAT in crude oils under pressure will be presented.

The Differential Scanning Calorimetry technique (using microDSC 7 or microSC) enables study of samples at controlled high pressure up to 1000bar and temperatures between – 45°C and 120°C. For experiments at lower temperatures (-196 to 200°C) BT 2.15 can be used.

431587 Calculation of Water Influx and Its Impact on Oil Recovery in Edge Water Drive Oil Reservoirs (Rapid Fire)

Abstract:

Natural influx of water in oil reservoirs surrounded by water bearing rocks (aquifers) play a very important role in increasing oil recovery. Such reservoirs are referred as water drive reservoirs. The pressure drop in reservoir caused by production of hydrocarbons prompts the influx of water from surrounding aquifer to offset it. This pushes more hydrocarbons towards the surface thus enhancing the oil recovery.

Correct prediction of amount of water influx is very important to find the recovery factor associated with a reservoir. Various models like Schilthius, Van Everdingen & hurt and Fetkowich have been proposed for water influx calculation. In this study attempts are made for using these models for calculating the water influx for a sample reservoir from Niger delta using the pressure history data. A MATLAB program is developed for this purpose. The effectiveness of each model is studied in explaining the behavior of given reservoir. The water influx calculated is used in material balance analysis to characterize the aquifer and find the value of initial oil in place (OOIP).

The water influx values obtained from VEH model and Fetkowich model are similar, with the only advantage of Fetkowich model being that it is computationally easier. The Schilthius model gives incorrect value of water influx for our case as the amount water influx given by Schilthius model does not give straight line when used in material balance analyses. The values of cumulative water influx obtained from VEH and Fetkowich model are in accordance with the year wise oil production data. Both year wise water influx and oil production reaches peak at around 4 years. This shows that the VEH & Fetkowich models are valid.

432661 Heat Stress Assessment on an Offshore Platform Using Computational Fluid Dynamics and Wet Bulb Globe Temperature Models

Abstract:

Heat Stress Assessment on an Offshore Platform using Computational Fluid Dynamics and Wet Bulb Globe Temperature Models

Kuochen Tsai, Shell International E&P

Paper #: **SPE-174963-MS**

Deep water oil discovery has played an increasingly critical role for the oil industry. With the increased water depth of new discoveries, oil temperature keeps hitting new highs as well and has huge impact for equipment and the work environment on the oil rig. A wet-bulb globe temperature (WGBT) model has been incorporated into computational fluid dynamics (CFD) software to assess heat stress for the safety of work environment on an offshore platform crowded with oil processing units. The model contains all critical elements needed to model WBGT, including conductive and convective heat transfer from the ambient air, the radiative heat from the hot processing unit surfaces, solar load and ambient atmosphere, and most importantly the cooling effect from perspiration evaporation. The feasibility of such a comprehensive model greatly enhances our understanding of the heat stress distribution and opened new opportunities for remedial actions.

The CFD model was constructed with geometry files directly imported from 3-D modeling software used for large offshore platform structures. The imported geometry was then extracted for necessary parts critical for the thermal modeling. Smaller features, such as deadleg pipes, nuts and bolts that won't influence the results are removed to reduce computational costs. The parts were then grouped according to thermal boundary conditions to be applied, including temperature, heat transfer coefficients and emissivity. A wrapping mesh was applied to convert the CAD geometry into a meshable domain and mesh was generated with specified sizes that can eliminate small gaps and undesirable features ill-suited for CFD solutions. A WBGT model was incorporated into the CFD software (ANSYS Fluent^{TM}) using an iterative procedure from the work of LilJegren et al. (2008). The inclusion of the WBGT model greatly enhances the predictive capability of CFD models for assessing the heat stress for human operators due to its ease of use without explicitly modeling the human body surfaces and locations.

The modeled results show that it is necessary to insulate the processing unit surfaces for safe operation due to the proximity to high temperature surfaces from the impact of thermal radiation. Other remedial measures were also explored and although the results were disappointing, the method opens up new possibilities that allows the evaluation of safe work environment in the planning phase and enables more effective project control.

References

J. C. Liljegren, R. A. Carhart, P. Lawday, S. Tschopp and R. Sharp, "Modeling the wet bulb globe temperature using standard meterological measurements," J. of Occupational and Environmental Hygiene, **5**, pp. 645-655, 2008.

433982 Pseudo-Component Modeling for Phase Behavior of Crude Oil and Water Under High Pressure and Temperature

Abstract:

We present a crude pseudo-component model based on Lee-Kesler correlations and predictive Peng-Robinson 1978 (PPR78) model to study the phase behavior of mixtures of crude oils and near-critical water or supercritical water. We develop a structural analysis of crude oils to bridge Lee-Kesler correlations and the PPR78 model, which enables the phase equilibrium calculations for the mixture of crude oils and water. This model is also applicable to study the phase behavior of mixtures of supercritical CO_2 and crude oils under the reservior conditions.

AUTHOR INDEX